Seed Health Testing: Principles and Protocols

Seed Health Testing: Principles and Protocols

Dr. Mahesh Kumar

RANDOM

RANDOM PUBLICATIONS

NEW DELHI (INDIA)

Seed Health Testing: Principles and Protocols

ISBN 978-93-5111-203-7

Published in 2014 in India by

RANDOM PUBLICATIONS

4376-A/4B, Gali Murari Lal, Ansari Road
New Delhi-110 002
Phone : +91-11-43580356, +91-11-23289044
e-mail: randomexports@gmail.com, sales@randompublications.com, info@randompublications.com

Reprinted 2021

Type Setting by : Keystoneprintads, Delhi-110051
Digitally Printed at: Replika Press Pvt. Ltd.

Preface

Seed health refers primarily to the presence or absence of disease-causing organisms, such as fungi, bacteria and viruses, and animal pests, such as eelworms and insects, but physiological conditions such as trace element deficiency may be involved. Seed health testing deals with detection and identification of seed-borne microorganisms. The objective of seed health testing is to determine the health status of a seed sample, and by inference of the seed lot, thus gaining information that can be used to compare the value of different seed lots. Seed health testing may elucidate seedling evaluation and causes of poor germination or field establishment and thus supplement germination testing.

The seed industry has a twofold responsibility in the area of seed health: to deliver sufficiently healthy seed to farmers and seed producers, and to respect international phytosanitary regulations. Seeds are the foundation to crop production and seed health is related to food production in various ways. Of particular concern to the seed industry is seed used for sowing. Seed-borne pathogens may cause disease or death of plants resulting in crop loss and thereby, food. In order to ascertain the health of commercial seed, seed health tests are required.

Seed borne viruses and pathogens can cause damage or death to crops, adversely affecting yields and marketability. Our trained scientists can help you to identify which plant disease(s) are affecting your crops, allowing you to make informed decisions about domestic sales and exports. With cutting edge technology and the latest testing methods we have the equipment and experience to meet all your seed health testing needs.

Many important diseases of pulses can be seed-borne. Pulse growers can minimize losses from these diseases by using high quality seed. Seed testing is required to establish whether seed is infected. Seed health tests are currently available to detect the most important seed-borne pathogens of pulses. Only seed that is pathogen-free should be used for sowing. Testing seed before sowing will identify potential disease problems and allow steps to be taken to reduce the disease risk. Laboratory testing is usually required, as infected seed may have no visible disease symptoms. The book details procedures of routine seed testing methods and recent approaches in seed health testing.

I thank all members of my team who have helped in the preparation of the book. My special thanks go to "Random Publications" who have published the book.

– *Dr. Mahesh Kumar*

Contents

1

Germination of Seed Testing

When germinating seed it is helpful to remember that a seed is made up of three parts:

- An outer protective coat
- A food supply under the seed coat (the endosperm)
- An embryo of a young plant.

The protective coat prevents sprouting until ideal growing conditions exist. Bringing seeds out of dormancy involves manipulating conditions to hasten germination. Even with ideal conditions, some seeds are still very difficult to germinate.

There are several factors that affect germination. Water (moisture), light (or dark), oxygen and heat play a part in triggering germination. In addition to environmental factors, seed must be viable. Water is essential in the first phase of germination. Water penetrates the seed coat and causes the endosperm to swell. The seed coat, softened by water, splits open as the endosperm swells. The water dissolves nutrients in the endosperm making them available to the embryo and growth begins.

The growing medium must be constantly moist, but not wet. Any dry period may cause death of the sprouting embryo. Light can stimulate or inhibit a seed's germination. This determines whether the seed should be sown on the surface of the growing medium or below the surface. Check the seed packet or catalogue for light requirements. Oxygen is required by the embryo to begin growing.

The seed must respire to break down the food stored in the seed. This is one reason for using a light, well-aerated growing medium for starting seeds. Every seed has an optimum temperature range for germination. Many seeds have a fairly wide temperature range for germination, but some are limited to a narrow range. The temperature range is usually given on the seed packet or in the catalogue. The temperatures required by many seeds are higher than those in most homes.

The desired constant temperature can be achieved through heating cables placed under germination containers. Setting flats or pots on radiators, the furnace or on the refrigerator will provide bottom heat. However, these locations may be too hot and cause the soil to dry too quickly. Once germination occurs, a different, usually lower, temperature may be required for optimal growth of the seedlings.

INTRODUCTION TO SEED

Seed is the starting point of agriculture and dictates ultimate productivity of other inputs. Good quality seed alone increases the yield by 15-20 per cent. To meet the potential challenge of catering to the food need of 1.4 billion people of our country by 2025, a quantum increase in agricultural productivity is very much essential and hence production and distribution of high quality seeds of improved varieties/hybrids to the fanning community is becoming increasingly important. The expansion of agriculture under tropical conditions due to the improvement of cultivars with juvenile period imposed a scientific and technological challenges concerning the seed production under different environmental conditions.

The seed programme includes the participation of state government, SAU system, public sector, and cooperative and private sector institutions. With the best efforts of all these organized sectors, only 15-20 per cent of the total requirement of quality seed is being met with. In most, kind of seeds, the farmers depend on their own farm saved seeds for crop production, which needs certain basic practices of selection of good seeds for sowing.

Moreover the crops are raised for market and a small portion of the grains are separated, stored and used as seeds in the next season which may not meet the quality aspects as expected for a seed which results in poor field stand, and ultimately yield.

Quality seed is the key input for realizing potential productivity. As the quality deteriorates during subsequent generations the old seed must be replaced with fresh lots of quality seeds. Ideally seed should be replaced every year for hybrids and every three to four years for non-hybrids. Therefore it is necessary to improve the availability of quality seeds to raise the Seed Replacement Rate (SRR). Despite implementation of the organized seed programme, the seed replacement rate has only reached 15 per cent and there exists an alarming gap between the demand and supply of quality seeds.

The role of private seed industry in the production and distribution of quality seeds is well recognized in the Indian' seed industry. However, they remain generally in the production of low volume high value seeds, which cater. to the needs of only few selected farmers.. The distribution of high volume low value seeds such as rice varieties, oil seeds and pulses are still with the public sector organization. The non-availability of quality seeds in oilseeds and pulses is one of the main reasons for its lower seed replacement rate.

The immediate increase in the productivity and production of these corps can be achieved by a higher distribution of quality seeds of new and high yielding varieties. There is vast scope to produce and distribute quality seeds in these corps for which seed village concept is a noval and highly practical approach and needs to be promoted to facilitate production and timely distribution of quality seeds of desired varieties at village level. In this context, the concept of seed village which advocates village self-sufficiency in production and distribution of quality seeds is getting momentum.

Seed technology is the science dealing with the methods of improving physical and genetical characteristics of seed. The various aspects coming under seed technology are seed production, seed processing, seed certification, seed testing, seed storage, seed biology, seed entomology, seed pathology and seed marketing.

CONCEPT

- Organizing seed production in cluster (or) compact area
- Replacing existing local varieties with new high yielding varieties.
- Increasing the seed production
- To meet the local demand, timely supply and reasonable cost
- Self sufficiency and self reliance of the village
- Increasing the seed replacement rate

FEATURES

- Seed is available at the door steps of farms at an appropriate time
- Seed availability at affordable cost even lesser than market price
- Increased confidence among the farmers about the quality because of known source of production
- Producer and consumer are mutually benefited
- Facilitates fast spread of new cultivars of different kinds

INFORMATION

In many countries of the world the number of trees planted increases each year. A recent survey (Lanly 1982) has estimated that the area of forest plantations in tropical countries will increase from 11.5 to 17.0 million ha between 1980 and 1985, an increase of 48 per cent in five years. Of the 11.5 million hectares at the end of 1980, 40 per cent had been planted in the previous five years.

The average annual planting rate is expected to increase from 0.92 million ha in the period 1976 – 80 to 1.10 million ha in the period 1981–1985, an increase of 20 per cent. Large planting programmes are also in operation in many temperate countries. In addition to new planting, replanting of harvested plantations of non-coppicing species must be carried out on a considerable scale each year.

Large as these areas may appear, they amount to only about a tenth of the area of natural forest being destroyed in the tropics over the same period. A further increase in planting rates beyond 1985 will certainly be necessary. Forest plantations are a powerful tool in the continuing efforts of foresters to increase productivity per unit area—the only means of reconciling the increasing demands for forest products and services on the one hand with a decreasing area of land available for forestry on the other. A combination of intensive site preparation with the use of uniform, well-grown nursery stock, planted at uniform spacing, increases growth and yield, reduces rotation length, facilitates tending and harvesting operations and improves the quality and uniformity of wood, as compared with natural forest.

Plantations also offer the means of using on a large scale the genetically improved material developed by tree breeders. Although there is no case for the indiscriminate replacement of all natural forest by plantations, their judicious use, by providing an alternative source of forest products, can itself reduce pressure on the remaining natural forest and so help to conserve it as a habitat and a source of genetic diversity. Not only do plantations have a major role as producers of timber, pulpwood and wood-based panels for forest industries, but fuel wood and pole plantations and farm woodlots are locally important in many countries. Tree-planting is not confined to block plantations. Shelterbelts and dispersed planting for soil stabilization, habitat improvement, urban and rural amenity or as part of an agrisilvicultural system all benefit the human environment.

With such a variety in planting purposes, it is not surprising that the scale of tree-planting and the variety of species planted continue to grow in so many countries. The current greatly increased interest in agro forestry opens up a whole new range of species for trial. Ability to grow in symbiotic relationship with agricultural crops will be the essential characteristic and will involve criteria such as rooting habit, ability to fix nitrogen and multipurpose uses (food, wood, shelter). Low stature may be beneficial and shrubs may become as important as trees. These new developments will introduce new opportunities and new problems in seed collection and handling. With a few exceptions, notably among the poplars and willows and in some tropical species of Casuarina, trees are propagated from seed, and the suitability and quality of the seeds have a big effect on the success of the plantations raised from them. The use of sound seed from stands of high inherent quality is widely recognized as the best means of ensuring fast-grown and healthy plantations capable of yielding high quality wood. Seed quality comprises both genetic and physiological quality.

It must be stressed that "good seed" implies seed which is both of high viability and vigour and is genetically well suited to the site and to the purpose for which it is planted. Physiologically good seed may lead to successful establishment of a plantation but this is of little value if it is slow growing, ill adapted to the site or produces the wrong kind of wood because the provenance or genotype was incorrectly chosen. On the other hand, there is little point in producing genetically improved seed at an increased cost if it is killed by poor handling techniques and has to be replaced or supplemented by inferior seed in order to achieve planting targets. Good seed handling is an essential complement to genetic improvement. Quantity, as well as quality, of seed production is important. In natural stands variation in the quantity of seed produced affects the forester's decision as to the years in which to collect seed and the trees from which to collect it, as described in Chapter. More intensive management affords him the opportunity of stimulating heavy seed production in genetically superior crops through deliberate treatment, *e.g.* thinning in seed stands or a combination of initial spacing, irrigation, fertilization and thinning in seed orchards. Description of these methods is outside the scope of this book, since they are properly a part of seed stand and seed orchard management, but programmes of seed supply should include measures to improve production quantitatively, as well as improving the genetic quality of the seed and the efficiency of collection and handling.

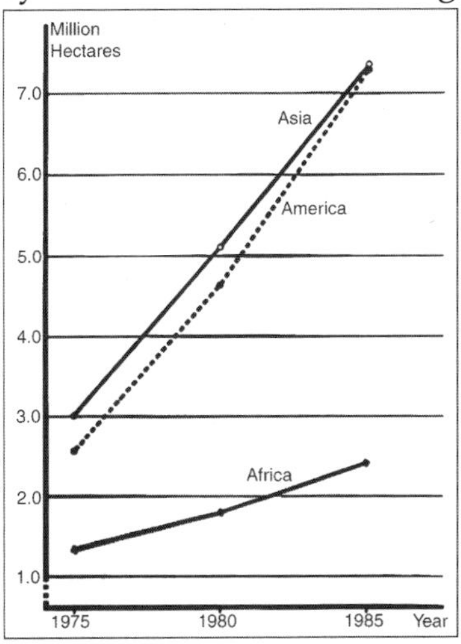

Fig. Estimated Total Areas of Forest Plantations in Tropical Countries

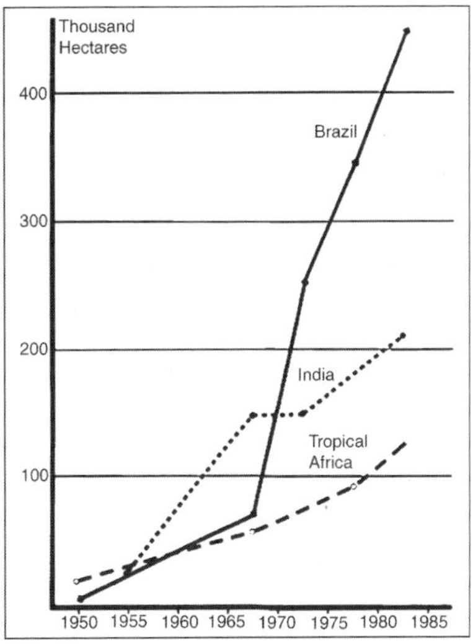

Fig. Estimated Annual Planting Rates Industrial and Non-industrial Forest Plantations.

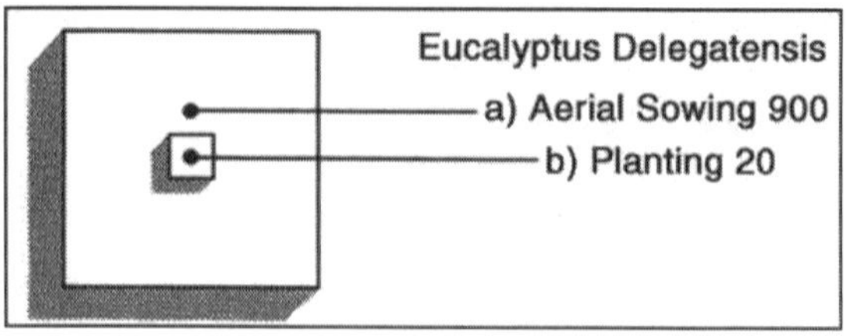

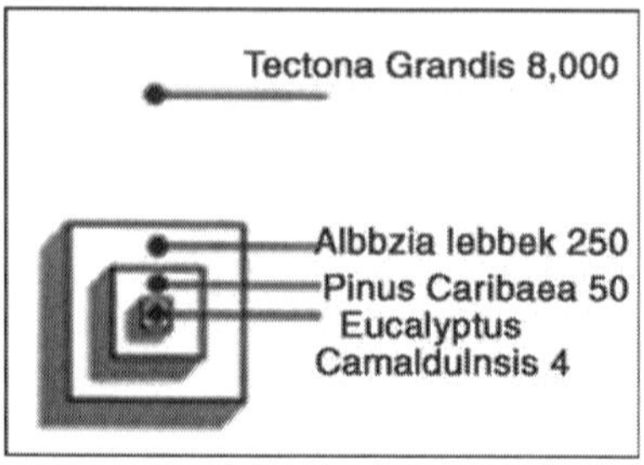

EFFECT OF SPECIES CHOICE ON QUANTITY OF SEED REQUIRED PER UNIT AREA

There is a considerable body of published information on seed handling in the temperate zones but published accounts of tropical experience with forest seeds are scattered and incomplete. While recapitulating the principles of seed handling derived from temperate experience, this guide seeks to illustrate these, as far as possible, by examples taken from tropical species. At the same time it must be emphasized that there is great variation in seed biology and that certain techniques which are commonly practised with good results in the temperate zones, *e.g.* stratification or prechilling, may have no useful function whatever in the tropics. It is dangerous to extrapolate temperate experience to tropical species, experience in the dry tropics to rain forest species, or experience in tropical agriculture to tropical forest tree seeds, without testing it in each case.

Valuable lists of equipment and suppliers have been published in the last decade. That published by ISTA (1982) covers equipment used in seed testing. The older list compiled by Bonner (1977) covers all operations from seed collection to seed testing and includes the names of equipment users as well as suppliers. Both equipment and techniques must be adapted to local species and conditions. For example hard-coated seeds of high natural longevity which are produced regularly every year require no more for storage than a well-ventilated room with protection against pests and diseases. A domestic refrigerator or "deep freeze" may be perfectly adequate for long-term storage if only small quantities are to be used each year. Large units and mechanized equipment may be essential in some countries, but heavy expenditure on such items should be incurred only after a full evaluation of the alternatives. The chain of seed handling operations is only as strong as its weakest link. Therefore great care must be taken at all stages if seed viability is to be maintained all the way from the parent tree to the nursery bed. If a seed has lost viability in an early stage of the process, the best storage or pretreatment can never bring it back to life.

Perfect extraction and cleaning of seeds are a waste of money if they are killed later by incorrect storage conditions or careless handling in transit. The greatest risks to seeds occur during temporary storage immediately after collection, during transit to the processing depot, and again during transit from the seed store to the nursery. These are the periods when sophisticated equipment is not available and everything depends on common sense and meticulous care to ensure good ventilation and avoid extremes of temperature. "International exchange of tree seeds is an opportunity to share the world's forest wealth" (Baldwin 1955). The widespread success of species such as Pseudotsuga menziesii, Pinus radiata, Eucalyptus globulus and Tectona grandis is a convincing demonstration of the advantages which many countries can derive from the introduction of exotics.

International cooperation in seed procurement is now practised by many countries and has long been supported by international organizations such as FAO and IUFRO. It should be increased. At the same time the movement of seed between countries does involve some additional problems in maintaining both the viability and the identity of the seeds. Phytosanitary regulations, including inspection and treatment, must be complied with in order to reduce to a minimum the risk of importing new and dangerous diseases. But every effort must be made to eliminate all unnecessary delays in transit, including those imposed by customs, plant hygiene and air lines, and to avoid excessive or duplicated fumigation treatments which kill the seeds.

Both sender and recipient must be aware of all the current regulations in force in the exporting and importing country and must plan the seed shipment well in advance to ensure smooth and unimpeded movement of the consignment. Apart from a few well-known species such as Tectona grandis, research on tropical forest seeds has been inadequate in comparison with both the severity of the problems and the large number of species of potential value for plantations. Much has still to be learned. A first step to this is a good understanding of the natural reproductive biology of each species. For species of the dry tropics which survive naturally by means of seedcoat dormancy, storage should present little problem and a modest programme of research should indicate the most appropriate pretreatment to overcome dormancy and induce uniform nursery germination.

The problem of preserving viability in recalcitrant seeds of tropical rain forest species, especially those which will not survive temperatures below about 10° C, is a much more intractable one. A variety of possibilities has been suggested (King and Roberts 1979) but so far little progress has been made towards a practical method of storage which would be applicable in large-scale afforestation projects. Much more research is needed to solve this problem. Until a solution is found, the dipterocarps and other recalcitrant species of the tropical rain forests will remain as much non-starters for widespread plantation forestry as they have been in the past.

SOWING SEEDS INDOORS

Sowing seed indoors is the easiest and cheapest method of producing vegetables, annual flowers and some perennial plants. Plants with extremely small seeds or those that need a long growing season make excellent candidates for starting indoors. Very small seed is difficult to plant outdoors because of size. Plants that require a long growing season may not have enough time to reach maturity unless started indoors in winter or early spring. Supplies needed for indoor seed sowing include the following: fluorescent or grow lights, disinfested containers with excellent drainage, pasteurized (sterile) seed-starting medium and a location with proper temperature and ventilation.

GROWING MEDIA

Choose a medium with a loose, uniform, fine texture. A pasteurized mixture that is 1/3 soil, 1/3 sand, vermiculite or perlite, and 1/3 peat moss has the qualities of a good seed-starting medium. Retail garden centres carry mixes labelled for seed starting. Whatever is selected, be sure it is pasteurized (sterile). Using pasteurized soil prevents damping-off, a fungal disease that kills young seedlings. Pasteurized soil also helps to avoid weeds, diseases and pests. Seed-starting media are usually low in fertility. This means that a regular fertilization programme is very important once seedlings emerge.

Containers

Any recycled containers are adequate for seed starting provided they are disinfested, have good drainage and are at least 2 inches deep. Other container options include compressed peat pellets, peat pots, paper pots, plastic cell packs and flats. Peat and clay containers tend to dry more quickly than plastic containers because they are very porous.

Sowing Seed

The correct timing of seed sowing is an important factor in successful indoor seed starting. In winter months, overanxious gardeners may sow seeds too soon. Seedlings that are held indoors too long perform poorly once transplanted into the garden. Most seeds should be sown 4 to 12 weeks prior to transplanting into the garden. The time it takes for seedlings to be ready for transplanting outdoors will vary. An ACCLIMATION period before placing seedlings directly into the permanent growing site must be included. However, readiness for outdoor planting will vary with how quickly germination occurs, the growth rate and weather conditions. Quality and quantity of light, temperature and nutrients affect the growth rate. Seed catalogues and packets provide information on days to germination and weeks needed to reach transplant size. Fill the container to within 1/4 inch of the top of the container with moistened seed-starting medium. After sowing seeds, keep the medium moist, not wet. To keep the medium moist, you may place the container in a plastic bag just large enough for the container. Seal the bag. The plastic bag keeps moisture in, but allows air exchange. The plastic bag method should not require any further watering until germination. Provide proper light and temperature conditions. Once seedlings germinate, remove the container from the plastic bag. Place the container in a location that has high light intensity and cooler temperatures.

Use the following rules to sow seed, depending upon seed size. Sow very small seeds by sprinkling on top of the medium and pressing in. Use a fine mist of water to gently wash seed into the growing medium. Sow medium-size and larger seed in rows 1 to 2 inches apart, and 1/8 to 1/4 inches deep. If no depth is specified on the seed packet, use the general rule of planting the seed at a depth twice the diameter of the seed.

When sowing in a tray or flat, sowing in rows is preferred over scattering seed. This method provides better air circulation than scattering of seed. When seedlings are crowded, they may become tall and spindly. To avoid the need to transplant seedlings from a seed flat to pots, you may sow seeds directly into cell packs or peat pots. Plant two or three seeds per cell or pot. When they germinate, remove the two less vigorous seedlings.

GERMINATION TESTING IN THE NURSERY

A good example of a suitable method for testing germination in the nursery is the procedure recommended for tropical pines in West Malaysia:

"Take 400 pure seed and divide into 4 samples each of 100 seed. Take 4 wooden or plastic boxes (opaque flexible plastic sandwich boxes are very good) of dimensions 30 × 30 cm and of depth about 10 cm and fill them slightly overfull with sieved sand to the top of the sides. Wet the sand by applying 0.5 litres of water to each box. Level the sand across the top of each box and make drills at 2.5 cm intervals and of depth 6 mm (for P. oocarpa the drills should only be 3 mm deep). Sow the seed (100) at intervals of 2.5 cm and gently press them on to the sand in the drill.

Cover to a depth of 6 mm (3 mm for P. oocarpa) with sand which has been passed through a Mesh No. 12 (12/64" or 4.76 mm) but retained on a Mesh No. 8 (Size 8/64" or 3.18 mm). Lightly water with a fine mist spray. Cover with transparent polythene sheeting of 0.2 mm thickness mounted on the upper side of a square tight-fitting wooden frame of 5 cm thickness. Within 24 hours condensation of moisture will appear on the inside of the polythene cover. If moisture disappears at any time during the ensuing 7 days then give a light mist spray and close lid once more. Germination will usually begin by the 7th day and the cover can be removed. Keep sand moist. A seed is recorded as having germinated when it has reached a height of 1 cm complete with seed coat enclosing the cotyledons. Record all germination from 7th to 28th day separately for each of the four boxes. Remove seedlings as soon as they are recorded plus any diseased ones which may not yet have reached 1 cm height as they may infect others. On the 28th day after sowing sieve the top layer of sand in each box through a No. 12 mesh screen (mesh size 12/64" or 4.76 mm) and record the number of full seeds which have not yet germinated. (Use cut test.)"

TESTING HOMOGENEITY OF GERMINATION RESULTS

The use of 4 replicates in a germination test enables the tester to measure the degree of variation in the sample. A simple method is to calculate the range of difference in germination per cent between the highest and the lowest of the sub-samples. The range can then be compared with the table published by ISTA, reproduced here as Table (it is applicable to replicates of equal numbers of seeds, not of equal weights). Provided that the actual range is less than the maximum in the table, the sample can be regarded as homogeneous and the average of the four replicates is accepted. If the actual range exceeds the maximum in the table, then a new sample must be drawn and tested. Common causes of such situations are non-homogeneity of seed and, in nursery tests, fungal or insect damage in one or more of the replicates. Retesting is also necessary when an extremely high percentage of full, ungerminated seeds are left at the end of the test (Bonner 1974). In this case retesting should be done with a different pretreatment to try to overcome dormancy and improve total germination.

Average Percentage Germination		Maximum Range	Average Percentage Germination		Maximum Range
1	2	3	1	2	3
99	2	5	87 to 88	13 to 14	13
98	3	6	84 to 86	15 to 17	14
97	4	7	81 to 83	18 to 20	15
96	5	8	78 to 80	21 to 23	16
95	6	9	73 to 77	24 to 28	17
93 to 94	7 to 8	10	67 to 72	29 to 34	18
91 to 92	9 to 10	11	56 to 66	35 to 45	19
89 to 90	11 to 12	12	51 to 55	46 to 50	20

This table indicates the maximum range (*i.e.*, difference between highest and lowest) in germination percentage tolerable between replicates, allowing for random sampling variation only at 0.025 probability. To find the maximum tolerated range in any case calculate the average percentage, to the nearest whole number, of the four replicates: if necessary, form 100-seed replicates by combining the sub-replicates of 50 or 25 seeds which were closest together in the germinator. Locate the average in column 1 or 2 of the table and read off the maximum tolerated range opposite in column 3.

COMBINING PURITY AND GERMINATION TESTS

Purity tests of most commercial seed lots of Eucalyptus are not made because it is difficult or impossible to separate the seed from the chaff of some species. Species in which seed and chaff are extremely similar in size, weight and colour include E. cloeziana, E. regnans and E. delegatensis. other small-seeded genera in which separation of pure seeds is difficult are Alnus, Betula, Populus and Salix. Even when separation is possible in these species, it is very time-consuming.

In the seed laboratory in Canberra it takes only 6–7 minutes to set up a test of eucalypt seed with 4 weighed replicates, but to carry out a purity separation and set up a 4 × 100 seed test takes from 20 to 50 minutes, depending on the degree of difficulty in separating out the chaff (Turnbull 1983). The small size of seed also precludes a cutting test to determine the number of full but ungerminated seeds at the end of a test. For these reasons tests are best made on replicates by weight and results recorded as numbers of germinated seeds per unit weight of the impure mixture of seed and inert material. A squash test may also be used to give a rough estimate of viability.

Where seed is sown broadcast, the practising forester is primarily interested in the number of plants he can expect to get from a given weight of the seed lot he receives. Provided he is told that 1 kg of "impure" seed should produce 36,000 germinated seedlings, he is not worried whether this is caused by a combination of 90 per cent purity × 80 per cent pure seed germination or 80 per cent purity × 90 per cent pure seed germination. For local tests, therefore, if seed laboratory staff is insuffient, there is no objection to omitting the purity test even on species for which it would be practicable. In the case of direct sowing into individual containers, however, there are obvious advantages in having separate figures for purity and germination, since sowing is by numbers of (one to several) seeds per container rather than weight of seed per m^2 of bed or tray.

INDIRECT TESTS OF VIABILITY

Estimating the germination potential of a seed lot by actually germinating a sample of it is often the method most relevant to practical forestry. But the tests take several weeks to complete and for some species pretreatment may take some additional weeks or months.

For this reason much research has been conducted to find other methods by which seed viability can be estimated accurately but much more rapidly than by germination testing. The following account of these methods closely follows that of Turnbull.

The objects of quick viability tests are:

- To determine quickly the viability of seeds of species which normally germinate slowly or show dormancy under the normal germination methods;

- To determine the viability of samples which at the end of the germination test reveal a high percentage of fresh ungerminated or hard seeds.

Only two methods, the topographical tetrazolium test and the embryo excision test were previously accepted by the International Seed Testing Association as official methods for some species of seeds. ISTA has recently accepted the X-ray method as a valid alternative to the cutting test for the detection of empty and insect-damaged seeds. The following tests can be applied depending on the circumstances:

CUTTING TEST

The simplest viability testing method is direct eye inspection of seeds which have been cut open with a knife or scalpel. If the endosperm is of normal colour with a well developed embryo, the seed has a good chance of germinating. This test is not very reliable. Seeds with milky, unfirm, mouldy, decayed, shrivelled or rancid-smelling embryos and abortive seeds that have no embryo can be judged as non-viable without much difficulty (Bonner 1974).

But it is not possible to distinguish moribund, recently dead or recently injured seeds which still appear the same as sound ones. The cutting test, as already mentioned, is used at the end of a germination test to determine the apparent viability of ungerminated seeds; it is also a useful tool in estimating the size and maturity of the seed crop before collection and the efficiency of methods used in processing. In the philippines good correlation has been found between cutting and germination tests in fairly large-seeded species such as Leucaena,Intsia bijuga and Lagerstroemia speciosa, but germination per cent was consistently 10–20 per cent less than the percentage of sound seeds on cutting test.

TOPOGRAPHICAL TETRAZOLIUM TEST

The tetrazolium method is only one of a number of biochemical tests which have been developed for seed testing. The various tests have been briefly reviewed by Moore. The tetrazolium test was introduced in 1942 by G. Lakon in Germany. In this method living cells are stained red by the reduction of a colourless tetrazolium salt to form a red formazan. The method emphasizes the need for a knowledge of the soundness of individual embryo parts for predicting the development of embryos into countable seedlings.

The testing procedure is described in detail in the ISTA Rules which approve the test for some species of hardwoods and conifers which germinate slowly by regular germination methods. Normal practice is to soak the seeds in water for about 20 hours, then cut or puncture the seed coat to facilitate entry of the 1 per cent aqueous solution of tetrazolium (TZ) and immerse the seeds in the dark for 48 hours. The process can be greatly speeded up by cutting through the seed at a distance of one third from the micropyle and placing it in a Vitascope vacuum machine for only half an hour. This method gives satisfactory results in Denmark but interpretation of results requires more experience in the operator than the method of seed immersion followed by excision of the stained embryo. The test is done on 4 replicates of 100 seeds each.

Justice states that, while the tetrazolium procedure is good in principle, its practical use in routine testing is limited by many problems, including: difficulty in staining of some seeds; necessity of cutting of dissecting seeds to permit observation of stained parts; poor agreement with results of germination tests in some cases, especially for seed of low germination capacity; lack of uniform interpretation of staining and difficulty in interpreting the significance of different degrees of staining; and an increase in man-hours required to test 400 seeds compared to regular germination tests.

The necessity for an experienced analyst for the successful use of this "common sense test" is admitted by Moore. There is little doubt that the test can be useful for testing the viability of certain species, providing trained staff are available to prepare the seeds and evaluate the results.

EXCISED EMBRYO TEST

By this method, the seeds are soaked for 1 – 4 days and the embryos are then excised from the seeds and placed on moist filter paper or blotter discs in petri dishes. The tests are placed in the light at a constant temperature of 20°C. The condition of the embryos is examined daily. Depending upon the species and lot differences, the tests can be terminated after only a few days, up to a maximum of 14 days, or as soon as distinct differentiation into viable and non-viable embryos can be made.

The excised embryo test is similar to germination tests in that it measures the quality of the seed by their actual germination. In addition it allows some measure of the embryo dormancy to be made, by counting those seeds which, although not growing normally, have grown slightly, remained firm and have kept their colour for the test period. The test is not valid for previously germinated seeds and must not be applied to samples which contain any dry germinated seeds. The success of the test requires considerable skill and experience in the operator and the ISTA rules restrict it to only a few species.

In a comprehensive study, Schubert compared the excised embryo method with the tetrazolium method for determining the viability of dormant tree seeds. He concluded that the tetrazolium method should receive preference over the embryo excision method but that improvements in the tetrazolium test should be made by providing for the use of bactericides and stronger reducing solutions to resolve doubts in weakly stained tissues.

RADIOGRAPHIC METHODS

Radiography was first used to determine seed quality over 70 years ago. The studies of Simak and Gustafsson highlighted the X-ray technique as a diagnostic method of tree seed analysis. The X-ray contrast method which uses various contrast or radiopaque agents was developed and applied successfully to species of Pinus and Picea. The X-ray method permits the detection of empty seeds, mechanical damage and abnormally developed internal seed structures, measurement of the thickness of the seedcoat and assessment of the seed viability when combined with a contrast agent.

The X-ray contrast method is based on the principle of semipermeability. When seeds are treated with a contrast agent, for example aqueous $BaCl_2$ or vaporous $CHCl_3$, their living tissues are able to prevent its entry due to their semi-permeability, but the dead tissues become impregnated. The impregnated tissues absorb X-radiation more intensively than the unimpregnated ones and thus appear lighter on the film than the unimpregnated ones. The contrast permits living and dead tissue to be located in the seed and an estimation of its viability. There are now possibilities of using non-toxic water, instead of toxic $BaCl_2$ or $CHCl_3$, as a contrast agent for testing seed viability.

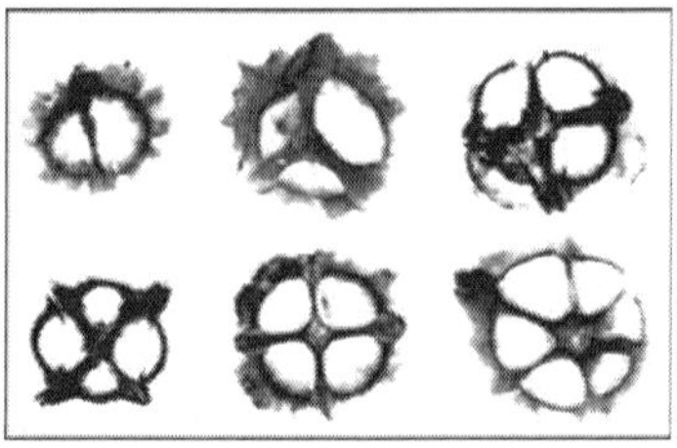

Fig. X-ray Radiograph of Teak Fruits Showing the Variation in the Number of Locules (two to six).

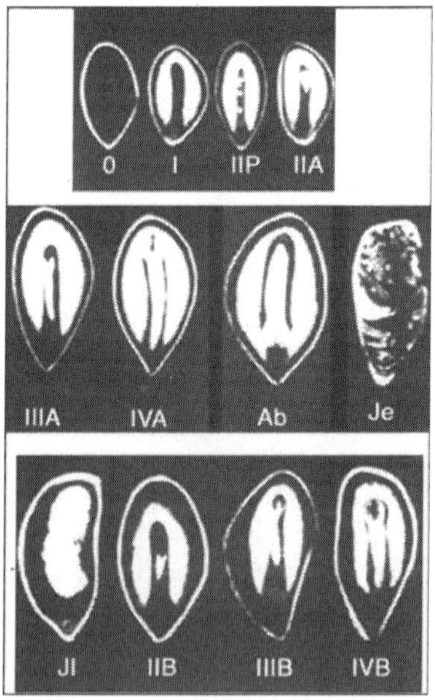

Fig. X-ray Radiographs Showing Embryo and Endosperm Classes in Coniferous Seeds.

O	Neither embryo nor endosperm (= empty seed)
I	Endosperm, embryo cavity developed but no embryo observed.
IIP	Endosperm, one or more small embryos the length of which does not exceed their breadth ("point embryos").
II	Endosperm, and one or several embryos, none of which is longer than half of the embryo cavity.
III	Endosperm, and one or more embryos, the longest of which measures between half and three quarters of the embryo cavity.
IV	Endosperm, with one fully developed embryo, completely or almost completely occupying the embryo cavity. Diminutive embryos rarely occur.
A	The endosperm almost fills the seed coat to capacity and easily absorbs the x-rays.
B	The endosperm fills the seed coat incompletely and is often shrunken or otherwise deformed. The x-ray absorption is inferior to that of class A.
Ab	Seed with abnormally developed endosperm or embryo.
J	Seeds damaged by insects, containing larvae (JI) or their excrement (Je).

The development of soft X-ray equipment has greatly simplified the operation. Complicated photographic equipment is not necessary and pictures can be made with polaroid film which provides clear and detailed radiographs within 30 seconds.

X-ray radiography has been successfully applied in determining the number of seeds in fruits of teak (Tectona grandis) and for studying their degrees of development. The technique has been tried on the fruits or seeds of sixty tropical forestry species and the results show that it can be reliably applied in processing such seeds.

A technique of stereoradiography as a supplement to the X-ray contrast method for use in seed quality testing has been developed by Kamra, Meyer and Wegelius. The chief advantage of stereoradiography is that it is possible for the observer to have a three-dimensional view of the object from a pair of radiographs. In this way, the exact topographical location of the contrast agent in the seed can be reliably determined. This increases the information which can be obtained from radiographs and adds to the analytical accuracy.

The X-ray method is a useful one and is likely to play an increasing role in seed testing. Earlier models of X-ray machines were expensive, but recent models, particularly from Japan are much cheaper and now cost less than a cabinet germinator. Improvements in photographic films and paper have speeded up the process and simplified the interpretation, so that technicians can be trained easily to produce consistent results. ISTA has accepted the method as a valid alternative to the cutting test for the detection of empty and insectdamaged seeds. It also shows considerable promise for distinguishing between viable and non-viable seeds among "full seeds".

For certain temperate conifers it has been possible to obtain good correlation between the Development Class (DC) of seeds, based on the development of both embryo and endosperm, and their germinability. The DCs which have been defined for conifers, and Table illustrates the germinability corresponding to each class, for Pinus sylvestris and Picea abies.

Table. Germinability of Fresh Collected, Undamaged Seeds (in per cent) Belonging to Different DCs, Selected by Radiography

Species	DC							
	I A	II A	III A	IV A	II P	II B	III B	IV B
Pinus sylvestris	0	50	88	99	0	5	43	68
Picea abies	0	36	82	97	0	15	71	92

HYDROGEN PEROXIDE

Hydrogen peroxide (H_2O_2) has a stimulating effect on seed germination and has been used in a rapid test for germination of several conifers in the western USA. Seeds are soaked overnight in 1 per cent H_2O_2. The seedcoat is then cut open to expose the radicle tip and the seeds put back into 1 per cent H_2O_2 in the dark at alternating temperatures (20° and 30°C). Counting and refreshment of H_2O_2 is done after 3 or 4 days and final assessment after 7 or 8 days. Radicle growth of 5 mm or more is scored "evident", 0 – 5 mm "slight" and no growth means a nonviable or empty seed. The test is quicker but less reliable than a normal germination test (usually producing a more rapid and higher final germination), slower but simpler to perform than excised embryo and easier to interpret than TZ.

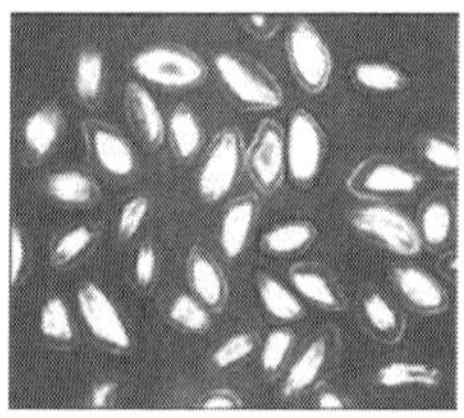

Fig. X-ray Radiograph of *Pinus caribaea* Seed. Most Seeds have Very Poorly Developed Gametophyte and Embryo and are Dead. A few Germinable are Indicated by the Black Outlines

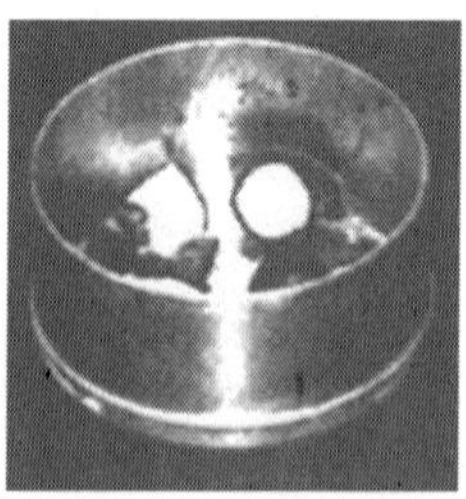

Fig. *Quercus* Seeds cut in Halves for oven Drying in Moisture Determination

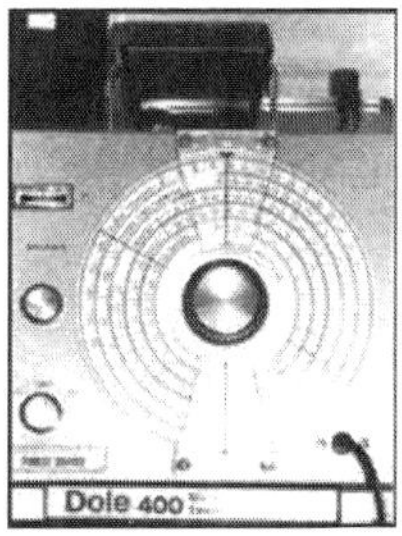

Fig. Dole Electric Seed Moisture Meter used in the USA

Fig. Electric Moisture Meters used in Denmark. A: Jacoby's in Frared. B: Super-matic. C: Mettler

TESTING MOISTURE CONTENT

The importance of moisture content in affecting the longevity of seeds in storage has been stressed in Chapter. In order to control the operations of drying (or moistening) of seeds in preparation for storage and to check the stability of moisture content during storage, it is clearly essential to have reliable methods of measuring the amount of moisture in a given sample.

Methods of determining moisture content of seeds have been classified by Justice into (a) basic methods in which the moisture is driven out of the seeds by heat and measured by the loss of weight of the original material, or the weight or volume of the condensed moisture, and (b) practical methods designed for rapid routine work and standardized against one or more of the basic methods. Probably all the moisture cannot be driven out of seeds without driving out small amounts of other volatile constituents or causing chemical changes in the material which would result in weight changes. In applying any method, therefore, it is necessary to adhere closely to the prescribed procedure in order that the results of all tests made by that method will be comparable.

Until recently ISTA prescribed three possible procedures:

1. Drying in an oven for 17 hours at 103°C
2. Drying in an oven for one to four hours at 130°C and
3. Toluene distillation. Method (2) is applicable only to certain agricultural seeds and method (3) Previously used for Abies, Cedrus, Fagus, Picea, Pinus and Tsuga, has now been eliminated because it had ceased to be used in practice (ISTA 1981 c). This leaves only method
   - The "Low constant temperature oven method" as applicable to forest trees.

The test should be made on two samples of about 5 g each, drawn from the working sample including impurities, not on pure seeds. Large seeds should be ground, broken or cut into small fragments to facilitate drying and a good rule of thumb is that any seeds that average over 10 mm in diameter or length should be broken. The samples should be weighed and placed in metal containers, well-spaced to facilitate air circulation, within an oven which is maintained at a temperature of 103° ± 2°C for 17 ± 1 hours. At the end of that period the seed should be placed in a desiccator to cool for 30 – 45 minutes and then reweighed. The relative humidity in the laboratory where the final weighing is done should be less than 70 per cent, to avoid rapid re-absorption of moisture.

The difference in MC of the two samples should not exceed a stated tolerance per cent. If it does, a further pair of samples should be tested; otherwise the mean of the two samples is the final result. The tolerance earlier prescribed by ISTA for all species was 0.2 per cent but, as pointed out by Gordon (1979) and Bonner (1981), a single tolerance figure is not applicable to all species. At the 1983 ISTA Congress in Ottawa, the tolerances approved for moisture tests in tree seeds were agreed as follows:

Sample Condition	**Tolerance%**
Small seeds, moisture <12%, *e.g.*, Picea, Alnus	0.3
Large seeds, moisture <12%, *e.g.*, Carya	0.4
Small seeds, moisture >12%	0.5
Large seeds, moisture 12 to 25%	0.8
Large Seeds, moisture > 25%, *e.g.*, Quercus	2.5

For tropical tree seed laboratories wishing to conform to ISTA rules, this relaxation of tolerances will be a considerable help.

The calculation of moisture content should be made on a wet weight or fresh weight basis i.e,

$$\text{Moisture Content\%} = \frac{\text{Original weight} - \text{Oven weight}}{\text{Original weight}} \times 100$$

Although wet weight basis is prescribed by ISTA and is becoming increasingly the standard form for expressing moisture content, it is not yet universal. To avoid any doubt, the method of calculating moisture content should be stated explicitly on any certificate or statement of results.

As explained by Gordon and Rowe, provided that the initial fresh weight of a seed lot is measured and the initial moisture content (wet weight basis) calculated by oven-drying a sample, any new MC reached as a result of drying (or wetting) can be calculated directly from the new weight of the seed lot; there is no need for further oven-drying of samples at the new MC. The desired weight of the seed lot to be achieved through drying (or wetting) can be calculated by multiplying its initial weight by the initial dry matter percentage and dividing by the desired dry matter percentage. *e.g.*,

1. If initial wet weight of a seed lot = 50 kg and the MC (w e t weight basis), determined by oven-drying a sample, is 25 per cent, the oven-dry weight = 75 per cent of wet weight = 37.5 kg.
2. If a period of drying reduces the wet weight to 46.5 kg, the new MC

$$(\text{wet weight basis}) = \frac{(46.5-37.5)}{46.5} = \frac{9}{46.5} = 19.4\%$$

3. If it is desired to reduce the MC (wet weight basis) to 10 per cent, then desired oven-dry weight will be 90 per cent of new wet weight and the seed lot must be further dried until its wet weight

$$\text{must be further dried unit its wet} = \frac{50 \times 75}{90} = 41.67\text{kg.}$$

Electric moisture meters give rapid estimates of seed moisture but they are not considered accurate enough for official seed testing. Their rapid operation does make them very useful in certain situations, for example they should be accurate enough to check tree seed moisture as a guide to drying seeds for storage. Meter readings are converted to seed moisture content by means of charts supplied by the manufacturer or developed from calibration curves in the laboratory for the species in question. Most meters will not measure moisture above 15 – 20 per cent and require a minimum of 90 – 100 g of seeds for a test (Bonner 1981). A locally made, cheap and portable electric moisture meter has been used successfully in Thailand for several years to measure the MC of rice grains and could be used for tree seed of similar size. It measures electric capacitance and uses a 9-volt battery as the source of power.

Electric moisture meters are well suited to small seeds, but cannot be used for large seeds such as Juglans or Quercus; winged seeds such as Fraxinus are also difficult to measure. Large or winged seeds can be rapidly dried in a microwave oven. If the oven is preheated, drying can be completed in 5 minutes and weighing in 6 minutes immediately afterwards if in an electronic balance, or after 30 – 45 minutes' cooling in a desiccator if weighing is done in an ordinary balance. Results can be expected to be within 7 per cent at a probability of 0.05 in the case of large seeds of high MC such as Quercus, and within 2 per cent in Fraxinus and Carya, as compared with the more accurate results of the slower conventional methods.

A simple and cheap method of drying seeds quickly is to use an infra-red lamp (Gordon and Rowe 1982). A weighed sample is subjected to heat from an infra-red lamp with such an intensity that it loses all its moisture, without being burnt, in about 20 minutes. When the loss of weight ceases, the new weight is measured and the percentage loss calculated. An up-to-date account of the measurement of tree seed moisture content has been published recently.

COMPATIBILITY BETWEEN SEED TESTING STATIONS

To achieve compatibility of germination test results between different laboratories ISTA and AOSA co-ordinate and produce agreements between seed testing laboratories on the procedures to be applied for each species. We have discussed in Chapters how the proportion of seeds which will germinate in a germination test can be affected by the test environment.

Thus if seed testing laboratories used very different germination test conditions the results obtained in different laboratories could differ substantially. To aid compatibility ISTA and AOSA seek agreement between laboratories as to the most appropriate environment for testing the germination of a given species. In theory this implies agreement upon a single germination test environment. In practice alternative environments are often specified for a given species. This is either because of disagreement between laboratories as to the single most appropriate environment, or because of the need to provide alternative procedures for certain seed lots, or because of practical difficulties in providing certain environments (for example, alternating temperatures) in some laboratories.

The rules of the ISTA and AOSA list the prescribed germination test conditions for each species. These prescriptions specify the substrata, temperatures and test duration which seed testing laboratories must adhere to, Additional directions are provided for special treatments which may be required for certain species together with recommendations for breaking dormancy. These prescriptions and directions are summarised in the chapters which follow. Now, these rules and directions are intended to give the most regular, rapid and complete germination for the majority of samples of seed of a particular species. To that end we presume the rules are successful, but how closely does the above aim coincide with the problems facing those testing seeds within gene banks?

PRESCRIPTIONS ARE NOT AVAILABLE FOR ALL SPECIES

First, ISTA and AOSA tend to only provide rules for those species in which certian individual seed testing stations have knowledge of the germination behaviour of a large number of seed samples. Thus the species listed in the ISTA and AOSA rules are those in which there is considerable trade in the sees; the seed-propagated crops. In contrast gene banks will often need to germinate seeds of species which are not yet in substantial trade in the developed world (*e.g.* the tropical pasture species), true seeds of vegetatively propagated crops (*e.g.* potato, Solanum tuberosum), and seeds of species which do not have crop status such as wild and weedy relatives of modern crop species (*e.g.* teosinte, Euchlaena mexicana). Thus the first drawback for gene banks of the ISTA/AOSA rules is that for many species on which gene banks require advice on germination test conditions no information is provided.

PRESCRIPTIONS DEVELOPED FOR MODERN CULTIVARS

Secondly, within those species for which ISTA and AOSA prescribe test conditions the range of seed lots which seed testing stations test for germination, and upon the results of which ISTA precriptions are based, are generally limited to relatively non-dormant, high quality, modern, genetically homogeneous cultivars. That is in most cases the cultivars with which the rules have been developed are likely to have been subjected to considerable selection pressures. In contrast gene banks will be testing seeds of primitive cultivars and genetically heterogeneous landraces which may also show considerable dormancy and/or be of poor quality. Thus the second drawback for gene banks of the ISTA/AOSA rules is that the prescribed conditions for a particula species will not necessarily be suitable for those acessions within gene banks which are dissimilar to modern cultivars-or of poor quality.

RELATIVE PRIORITIES OF THE AIMS OF SEED TESTING

Thirdly, for some crops seed testing stations may operate under considerable pressure at harvest time. For example, in temperate countries autumn sown cereal seeds are sown only five or six weeks after harvest. Thus seed testers may need to complete germination tests in as short a time as possible. Hence seed testing stations' aim of providing conditions which give rapid germination. Now in at lest one group of species—the temperate cereals—although the conditions prescribed by the rules may give rapid germination this is only achieved in our experience by using supra-optimal temperatures at the cost of the parallel aims of achieving regular and complete germination which are better achieved at much lower temperatures.

In contrast to seed testing stations, gene banks are not under any pressure to give priority to rapid germination over achieving regular and complete germination. Thus the third drawback for gene banks of the ISTA/AOSA rules is that the aims of rapid and complete germination may not always be compatible. Note that this criticism does not apply to the rules for all crops. In the case of tree species, for example, considerable periods are allowed for dormancy-breaking and germination test treatments by ISTA and AOSA rules, and the requirement for complete germination takes precedence over the requirement for rapid germination.

FIELD PLANTING VALUE

The purpose of commercial testing is to provide information on the comparative field planting value of different seed lots. In general the ISTA and AOSA prescriptions make no attempt to enable those viable seeds which will not germinate under field conditions to germinate in the laboratory test. If, for example, dormancy is a problem in laboratory tests, but not in field sowings (for example, due to differences in temperature between the two regimes) then directions to break dormancy must be provided in the rules—otherwise the field planting value may be underestimated. However, where dormancy is a problem in both laboratory tests and field sowings then to obtain an indication of field planting value the rules may not necessarily provide dormancy-breaking directions—since the removal of dormancy in laboratory tests may result in an overestimate of field planting value.

Again whether or not this objective of ISTA or AOSA prescriptions and recommendations makes them unsuitable for viability testing in gene banks depends on the species. For example, in tree species directions for dormancy removal are suggested to enable viability to be estimated in laboratory tests and these directions must also be followed in field sowings if laboratory results are to provide a reasonable estimate of field planting value. In contrast in some species where hardseededness prevents germination no recommendation is made to render the seed coat permeable.

Instead the proportion of hard seeds is reported. Thus the fourth drawback of ISTA/AOSA rules is that no attempt may be made to overcome a wide range of physiological phenomena which may prevent the germination of viable seeds under both laboratory and field conditions. Of these phenomena the most important are hardseededness and strong dormancy. In contrast to seed testing stations, gene banks aim to provide conditions which will enable all viable seeds to germinate both in laboratory tests and in subsequent field or glasshouse sowings.

OTHER SOURCES OF PRESCRIPTIONS

Numerous groups of workers have attempted to provide prescriptions for germination tests of species for which no ISTA/AOSA rules are available or which overcome certain of the deficiencies of the ISTA and AOSA prescriptions for species where rules are available. Particular emphasis has been placed on overcoming dormancy since this can be an important problem for plant breeders, and also for seed testers if considerable dormancy is experienced when the seeds are tested but which is likely to be lost during storage before sowing.

However, much of this information is also of limited use in gene banks. This is because the majority of such information relates to the application of single dormancy breaking agents alone. It was explained that if this approach is used then the treatment concentration (if a chemical agent) or duration (if another form of treatment) required to break dormancy in the most dormant seeds would be likely to damage less dormant seeds, particularly if these are also of poor quality.

THE AIM OF DEVELOPING A SPECIES PRESCRIPTION FOR USE IN GENE BANKS

We believe that the ultimate desirable aim of a prescriptive test for a given species is to be able to provide within one test environment or procedure those conditions which enable all viable seeds of that species—whether strongly dormant, moderately dormant, weakly dormant or non-dormant, and of high or low quality—to be able to germinate. This means that any stimulus or combination of stimuli used to promote the germination of dormant seeds must not be damaging to other non-dormant seeds in the sample or to intolerant genotypes.

DEVELOPING A PRESCRIPTION FOR A SPECIES

Whether this aim can in fact always be achieved is a matter for debate, but two considerations are important in attempting to achieve it. The first is to ensure that the germination test temperature is suitable for poor quality seeds; within this range attempts can then be made to determine the most suitable temperature regime which will minimise the expression of dormancy. In certain species this first step may result in an adequate single germination test environment without any need for further treatment. In temperate cereals this appears to be the case, where a single low constant temperature is appropriate for both purposes.

However, in other species simple solutions of this type may not be available and so it is necessary to take a second step and provide further stimuli to break dormancy which do not impair the germination of less dormant seeds. In our opinion this is often best attempted by combining more than one additional stimulus with all stimuli being applied at relatively low dose-levels. This is an attempt to utilise the positive interactions observed between many dormancy-breaking agents. The treatments developed for Oryza spp. and Vitis spp. provide examples of this type of procedure (these and other prescriptions are described in the appropriate chapters of this volume).

Gene banks considering developing their own germina-tion test prescriptions may be discouraged by the wide range of different treatments—and the even wider range of different treatment combinations—which may be of potential benefit in promoting the germination of viable seeds. Consequently it is worthwhile considering which factors should receive priority in any investigation. We believe that the three most important factors to consider first are to ensure that the seeds have imbibed moisture without damage and then to attempt to determine the most suitable temperature (probably alternating temperature, and light regimes for germination.

Provided a prescription is already available for a species which is suitable for application in a gene bank then this would be the simplest approach for gene banks to use. However, even where a prescription is well-tried, gene banks will still need to be alert to the possibility that some accessions may be discovered for which the prescription does not appear to work. Furthermore, for many years to come, there will continue to be a large number of species for which there are no adequate prescriptions. Accordingly alternative approaches to providing suitable germination test environments are also considered in this handbook.

USE OF TETRAZOLIUM TEST TO DETERMINE EFFICACY OF A PRESCRIPTION

Although a gene bank can never be absolutely certain about the efficacy of any approach to germinating seeds, do not forget that the efficacy of a species prescription on a particular accession can be tested by comparing the results of the germination test with that of the topographical tetrazolium test on a sub-sample of the seed lot and by also applying a tetrazolium test to the seeds remaining ungerminated at the end of the test. In this way one can determine the proportion (if any) of viable seeds which are either killed by the germination test procedure or remain dormant at the end of the germination test procedure.

In the above we have detailed objections to the wholesale and unquestioning adoption of either ISTA or AOSA prescribed germination test procedures as gene bank germination tests. However, this certainly does not mean that such prescriptions should be ignored by gene banks. The ISTA and AOSA rules summarise a wealth and depth of expertise and experience which is unmatched. Thus where available, the prescribed conditions represent a control against which any proposed alternative should be tested; and, despite our earlier comments, certain of these prescriptions will be suitable for gene bank use.

IMPROVING SEED GERMINATION

When we want to produce large amounts of plants or simply when we want to start our gardens fast and get the most out of our purchases improving seed germination becomes a large priority. One of the largest concerns of world agriculture as well as the home grower is the decrease in germination time and increase in germination percentage since both of these factors can bring great benefits. Some seeds – especially some flowers and herbs – are often quite difficult to germinate and using certain techniques to increase the rate and speed in which they sprout has been the focus of a large amount of scientific research. On today's article I will be discussing the use of priming to decrease germination time, especially what priming is, what types are available and which ones you can use to decrease the germination time of those very difficult seeds.

To understand the concept of priming we first need a good grasp at the general concept of seed germination. A seed is a dormant embryo which carries within it the potential for a new plant's life. The seed is alive, yet has a very slow metabolic rate due to the low mobility of substances within the embryo's cells. This low metabolic rate allows the seed to remain alive, yet survive extremely long periods of time (some seeds can survive even hundreds of years) before actually sprouting into new plants.

GERMINATION

Is the process in which we awaken the embryo – increases seed metabolism and toggles the massive reproduction that causes a new plant to grow. The main mechanism that triggers this process is simply liquid water. When water gets into the embryo and hydrates its cells, it speeds up metabolism and allows the process of cell division and growth to rapidly increase. However it is not always this simple to start this process since several impairments – both chemical and physical – can exist for successful germination. Priming is simply a process done prior to conventional seed germination which allows the inhibiting mechanism to be broken and the metabolic speed increase to begin. There are several types of priming that can be done. A seed can be submerged in simple water (hydropriming), it can be soaked in a solution of a simple salt (halopriming) or it can be set in a non-ionic solution with high osmotic pressure (osmopriming). It is not entirely well established why one technique might work better than another but certainly some species tend to respond much more efficiently to one or another.

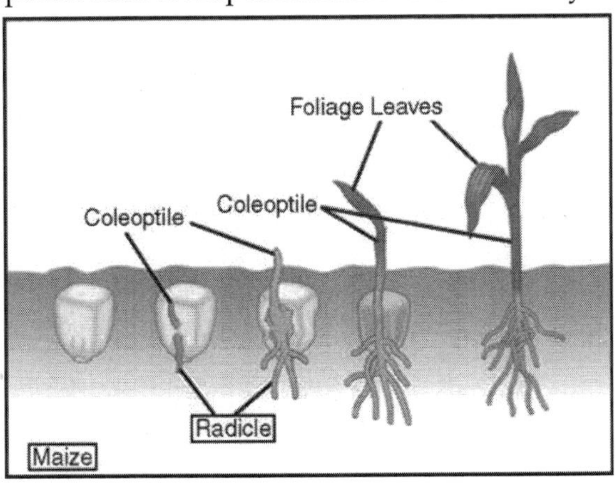

In general, priming offers the opportunity to almost always germinate seeds at much higher speeds without detrimental effects in germination percentages. For example, a two day treatment of parsley seeds with a PEG 6000 (PolyEthyleneGlycol) solution can reduce germination times substantially, from a few weeks to just a few days. Other seeds such as coriander might also benefit from similar treatments with PEG or with treatments with NaCl solutions. In general if you are looking to test priming on some difficult seeds you own you can try three small experiments to know which one works best for your particular seed variety and germination conditions. Do one experiment in which the seeds are simply soaked in water for 24 hours, another in which seeds are placed in a 200mg/L NaCl solution and another one in which the plants are submerged in a PEG 6000 20 per cent solution, then let the seeds air-dry after the treatments. After comparing the results of these experiments with a control with no priming you will be able to see which priming technique is better for you and most effectively increases your seed germination rates.

To sum it up priming your seeds is a very efficient technique to increase the speed of germination without sacrificing germination rates. This methods are not very useful for seeds such as lettuce or tomato – which germinate easily – but they are invaluable for plants such as parsley, coriander or carrots which are generally much harder to germinate. If you have some seeds that have been giving you a hard time or seem to take ages to germinate then setting up some priming experiments might be the best thing to do. Teachers may be unfamiliar with the concept of "priming" seeds. However, teachers understand the value of Early Childhood Education ... and how it can influence later classroom experiences in the higher grades! Giving a child a good start in education is important; the same can be said for seeds. Priming seeds is like giving them a "better start" to life.

Another way of explaining this process has been developed by Keith Kubik, Ph.D. a Seed Physiologist with the Harris Moran Seed Company in California. Dr. Kubik compared seed priming to building a pre-fabricated house. Pre-fabricated houses are built Inside A Factory. The walls and roof are put together in an environment that is ideal for the workers – very important in our climate here in Canada - and building efficiencies are best. However, the house is not completed inside the factory. The pre-built walls and roof are taken to the building site where the remaining construction takes place – the electrical work, the plumbing, and the basic fixtures like countertops and inside trim. Less time is consumed at the site, and the houses go up more quickly and more efficiently. Similarly, with seeds, the priming process starts the germination process in the laboratory. The basic chemical reactions or framework needed for the seed to germinate take place under ideal temperatures and high moisture conditions. The seed moisture is set so that the germination is ALMOST complete ... just like the prefabricated house when it is moved onto the actual home site. When you get your seeds, one of the two groups will be a control group and the other group will have been "primed" – almost ready for germination. Like the pre-fabricated house, seed germination of the primed seeds will probably take less time, because part of the germination process is already complete. You will still report on the germination process – the number of seeds planted and number of seeds germinated.

CLIMATE CHANGE IN FOOD PRODUCTION

Climate change may seriously threaten food production required for future generations across the globe. Climate change has many facets, for instance changes in long-term trends in temperature and rainfall regimes as well as increasing variability in extreme events; all affecting agricultural production. Wheat is one of the most important cereal crops; contributing about 21 per cent of the world's food. It is grown on about 200 million hectares of farmland worldwide. Changes in global warming are predicted to increase the frequency and severity of 'heat waves' in wheat growing areas in the world.

Heat and drought are the main abiotic constraints limiting the grain yield of cereals. In major wheat growing regions, temperature exceeding 30°C during reproductive stages reduces grain yield and quality (Maestri *et al*. 2002), which may be due to a wide range of interlinked processes, including accelerated development. Heat stress to varying degrees at different phenological stages, may cause a substantial yield reduction, primarily because of acceleration of vegetative growth, reduced duration of phasic developmental stages and carbon starvation owing to reduced net assimilation. The probability of heat stress around flowering, which can result in considerable yield loss, is predicted to increase significantly in coming years. As the world water supply is declining, drought is also threatening the world food security. Drought-induced losses in crop yield perhaps exceeds the loss from all other causes, as both severity and duration of stress are critical.

Seed priming is a controlled hydration process followed by re-drying that allows pre-germination metabolic activities, to proceed rapidly. Seed priming techniques have been employed to improve resistance against several abiotic stresses in a range of field crops. In this study potential of two seed priming techniques in improving wheat performance in conditions of changing climate was evaluated.

METHODOLOGY

Seeds of wheat cultivar SH-2002, used in this study were obtained from Wheat Research Institute, Faisalabad, Pakistan. Priming was achieved by soaked in aerated solutions of CaCl2 (-1.25MPa) and ascorbate (2 mM) for 12 h; while water soaked and dry seeds were taken as controls. Plants were raised under normal conditions (at 22/11°C) in a growth chamber till 3 leaf stage; after which stress treatments were imposed. After that the plants were divided into four sets *viz.*, 'optimal' conditions (100 per cent of field capacity at 22/11°C), drought stress (50 per cent of field capacity at 22/11°C), heat stress (100 per cent of field capacity at 26/13°C) and both drought and heat stress (50 per cent of field capacity at 26/13°C). The experiment was conducted in randomised complete block design with four replications. Data on relative leaf water contents, membrane permeability and plant biomass was recorded one week after the imposition of stress following protocol described elsewhere (Farooq *et al.*, 2008b). Data were statistically analysed by analysis of variance (ANOVA) techniques using compute software MSTATC. Graphical presentation of the data and computation of standard error bars was computed was made in Microsoft Excel.

RESULTS

Substantial reduction in plant biomass and relative leaf water contents was observed in wheat plants exposed to drought and heat stresses; none the less the extent of reduction was magnified when both the stresses were combined. Likewise the membrane stability was disturbed as indicated by higher electrolyte leakage. However seed priming techniques effectively alleviated the effect of drought and heat stress; osmopriming (with $CaCl_2$) being the most effective followed by ascorbate priming. Nevertheless the extent of improvement was higher under heat stress.

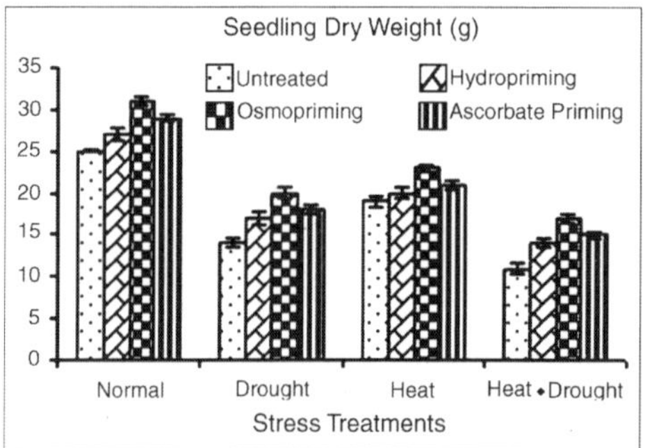

Fig. Influence of Seed Priming Techniques on Seedling Dry Weight in Wheat Under Drought and Heat Stresses Discussion.

The study indicates that wheat performance will be strongly influenced by heat and drought stresses, the major climatic threats. Nonetheless seed priming treatments may be employed to mitigate the stress-driven effects. Moreover, seed priming may improve the wheat performance under normal conditions as well.

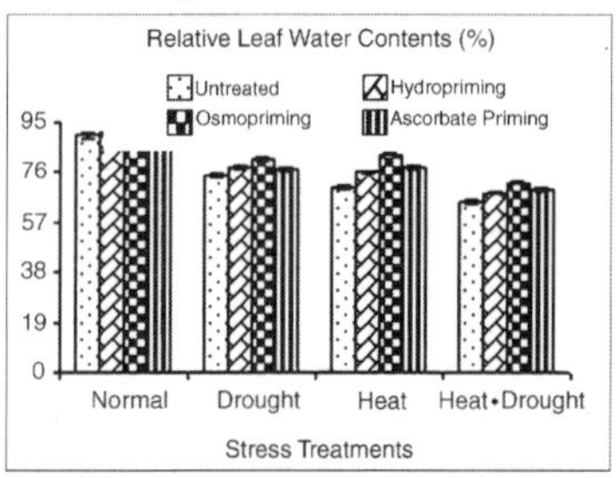

Fig. Influence of Seed Priming Techniques on Relative Leaf Water Contents in Wheat Under Drought and Heat Stresses.

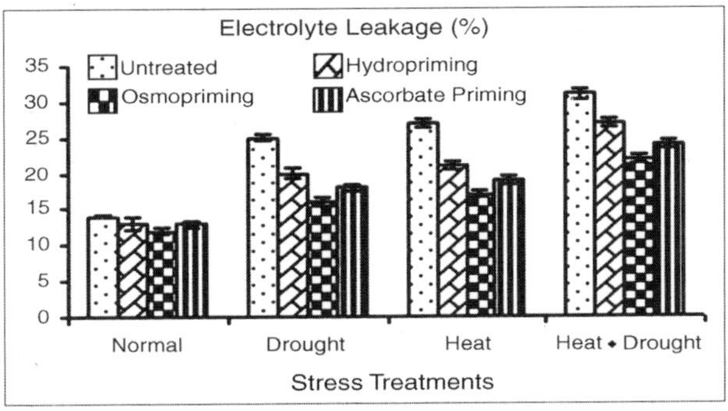

Fig. Influence of Seed Priming Techniques on Leaf Electrolyte Leakage in Wheat Under Drought and Heat Stresses.

Reduction in seedling biomass upon exposure to stresses may reflect altered leaf gas exchange attributes (data not shown) and/or tissue water status. Lipid peroxidation of membrane lipids is one prominent effects of abiotic stresses, resulting in disturbed membrane structure. Nonetheless seed priming treatments, in particular with $CaCl_2$, lessened the impact of the stresses assessed on seedling growth. Calcium, by maintaining the inter- and intra-molecular linkages, actually plays a pivotal role in the protection of membrane structures. In addition Ca^{2+} has been found to enhance the enzymatic antioxidants and also induce the heat shock protein synthesis. Several other reports also indicate the calcium mediated improvement in resistance against abiotic stresses.

2

Seed Born Diseases

Early blight is a very common disease of both potato and tomato. It causes leaf spots and tuber blight on potato, and leaf spots, fruit rot and stem lesions on tomato. The disease can occur over a wide range of climatic conditions and can be very destructive if left uncontrolled, often resulting in complete defoliation of plants. In contrast to the name, it rarely develops early, but usually appears on mature foliage.

SYMPTOMS

On leaves of both crops, the first symptoms usually appear on older leaves and consist of small, irregular, dark brown to black, dead spots ranging in size from a pinpoint to 1/2 inch in diameter. As the spots enlarge, concentric rings may form as a result of irregular growth patterns by the organism in the leaf tissue. This gives the lesion a characteristic "target-spot" or "bull's eye" appearance. There is often a narrow, yellow halo around each spot and lesions are usually bordered by veins. When spots are numerous, they may grow together, causing infected leaves to turn yellow and die. Usually the oldest leaves become infected first and they dry up and drop from the plant as the disease progresses up the main stem.

On tomato, stem infections can occur at any age resulting in small, dark, slightly sunken areas that enlarge to form circular or elongated spots with lighter-coloured centres. Concentric markings, similar to those on leaves, often develop on stem lesions. If infested seed are used to start tomato transplants, seedlings may damp off soon after emergence. When large lesions develop at the ground line on stems of transplants or seedlings, the plants may become girdled, a condition known as "collar rot." Such plants may die when set in the field or, if stems are weakened, may break over early in the season. Some plants may survive with reduced root systems if portions of stems above the canker develop roots where they contact the soil. Such plants, however, usually produce few or no fruits.

Stem lesions are much less common and destructive on potato. Blossom drop and spotting of fruit stems, along with loss of young fruit, may occur when early blight attacks tomatoes in the flowering stage. On older fruits, early blight causes dark, leathery sunken spots, usually at the point of stem attachment. These spots may enlarge to involve the entire upper portion of the fruit, often showing concentric markings like those on leaves. Affected areas may be covered with velvety black masses of spores. Fruits can also be infected in the green or ripe stage through growth cracks and other wounds. Infected fruits often drop before they reach maturity.

On potato tubers, early blight results in surface lesions that appear a little darker than adjacent healthy skin. Lesions are usually slightly sunken, circular or irregular, and vary in size up to 3/4 inch in diameter. There is usually a well defined and sometimes slightly raised margin between healthy and diseased tissue. Internally, the tissue shows a brown to black corky, dry rot, usually not more than 1/4 to 3/8 inch deep.

Causal Organism

Early blight is caused by the fungus, Alternaria solani, which survives in infected leaf or stem tissues on or in the soil. This fungus is universally present in fields where these crops have been grown. It can also be carried on tomato seed and in potato tubers. Spores form on infested plant debris at the soil surface or on active lesions over a fairly wide temperature range, especially under alternating wet and dry conditions. They are easily carried by air currents, windblown soil, splashing rain, and irrigation water.

Infection of susceptible leaf or stem tissues occurs in warm, humid weather with heavy dews or rain. Early blight can develop quite rapidly in mid to late season and is more severe when plants are stressed by poor nutrition, drought, or other pests. Infection of potato tubers occurs through natural openings on the skin or through injuries. Tubers may come in contact with spores during harvest and lesions may continue to develop in storage.

- Use a crop rotation that includes potatoes or tomatoes only every third or fourth year to allow infested plant debris to decompose in the soil. Rotations with small grains, corn or legumes are preferable.
- Use tillage practices such as fall plowing that bury all plant refuse.
- Select cultivars that have a lower susceptibility to early blight.
- Use certified disease-free tomato seed and transplants.
- If producing tomato transplants, disinfest soil in plant beds and control humidity in cold frames or greenhouses. Practice good sanitation throughout the transplant production operation.
- Use appropriate measures to control weeds and volunteer potatoes and tomatoes in production areas.
- Maintain fertility at optimal levels-nitrogen and phosphorus deficiency can increase susceptibility to early blight.
- Time applications of overhead irrigation to allow plants to dry before nightfall.
- Although the above measures are important to minimize infection, it is usually necessary to apply fungicide sprays to fully protect

plants from early blight. Applications to tomato are usually begun 2-3 weeks following emergence or soon after transplanting if a calendar schedule is followed. Timing of fungicides can also be made using the TOM-CAST system. For potatoes, fungicide applications should be initiated when plants begin to flower. Thorough coverage is important. Adjust equipment for good vine penetration.

LATE BLIGHT OF POTATO AND TOMATO

Late blight is one of the most devastating diseases of potato and tomato worldwide. It was responsible for the devastating Irish potato famine of the 1840's and has continued to be important to the present. Since 1990, late blight has caused widespread damage across the United States and Canada. If left unmanaged, this disease can result in complete destruction of potato or tomato crops.

SYMPTOMS

Late blight appears on potato or tomato leaves as pale green, water-soaked spots, often beginning at leaf tips or edges. The circular or irregular leaf lesions are often surrounded by a pale yellowish-green border that merges with healthy tissue. Lesions enlarge rapidly and turn dark brown to purplish-black. During periods of high humidity and leaf wetness, a cottony, white mold growth is usually visible on lower leaf surfaces at the edges of lesions. In dry weather, infected leaf tissues quickly dry up and the white mold growth disappears. Infected areas on stems appear brown to black and entire vines may be killed in a short time when moist weather persists.

On potato tubers, late blight appears as a shallow, coppery-brown, dry rot that spreads irregularly from the surface through the outer 1/8-1/2 inch or more of tissue. On tuber surfaces, lesions appear brown, dry, and sunken, while infected tissues immediately beneath the skin appear granular and tan to copper-brown. When tubers are stored under cool, dry conditions, lesion development is retarded and, upon prolonged storage, lesions may become slightly sunken and desiccated. Secondary bacteria and fungi frequently enter late-blight lesions, usually resulting in a slimy breakdown of entire tubers.

Late blight can also develop on green tomato fruit, resulting in large, firm, brown, leathery-appearing lesions, often concentrated on the sides or upper fruit surfaces. If conditions remain moist, abundant white mold growth will develop on the lesions and secondary soft-rot bacteria may follow, resulting in a slimy, wet rot of the entire fruit.

Causal Organism

Late blight is caused by the fungus *Phytophthora infestans*. Unlike most pathogenic fungi, the late blight fungus cannot survive in soil or dead plant debris. For an epidemic to begin in any one area, the fungus must survive the winter in potato tubers (culls, volunteers), be reintroduced on seed potatoes or tomato transplants, or live spores must blow in with rainstorms.

Disease development is favoured by cool, moist weather. Nights in the 50's and days in the 70's accompanied by rain, fog or heavy dew are ideal. Under these conditions, lesions may appear on leaves within 3-5 days of infection, followed by the white mold growth soon thereafter. Spores formed on the mold are spread readily by irrigation, rain and equipment. They are easily dislodged by wind and rain and can be blown into neighbouring fields within 5-10 miles or more, thus beginning another cycle of disease.

Infection of potato tubers arises from spores that develop on foliage. Tubers exposed by soil cracking or erosion of hills may come in contact with spores washed down from infected leaves and stems by rainfall or irrigation. Tubers infected during the growing season may partially decay before harvest. Tuber infection may also occur at harvest when tubers contact living spores remaining on infected vines. Little if any tuber-to-tuber spread of late blight occurs during storage if tubers are kept under cool, well-ventilated conditions.

The protectant fungicides commonly used to protect plants from late blight remain fully effective with all known strains of the fungus:

- Infected cull potatoes are a major source of spores of the late blight fungus and must be disposed of properly-Do Not Make Cull Piles. Cull potatoes should be spread on fields not intended for potato production the following year in time that they will totally freeze and be destroyed during the winter. If this is not possible, they must be destroyed in some other way such as by complete chopping, burial, burning or feeding to livestock.
- Plant only certified seed potatoes. Use of "year-out" seed or seed saved from local crops is asking for trouble with late blight. Seed sources should be selected very carefully to avoid bringing in late blight on seed potatoes, especially new strains of the fungus. Look for the characteristic coppery-brown discolouration of the potato flesh under the skin of seed tubers. Infected tomato transplants also can be a significant source of the disease. Use only obviously healthy tomato transplants free of dark lesions on leaves or stems.
- Volunteer potatoes and tomatoes can be a significant source of spores of the late blight fungus. All volunteers should be destroyed as quickly as possible by herbicides, chopping, or cultivation.
- Growers should scout fields regularly to look for late blight. Special attention should be paid to early-planted fields because that is where the disease is likely to develop first. Scouting should be concentrated in low-lying areas, field edges along creeks or ponds, near the centre of centre-pivot irrigation rigs, in areas near woodlots or any area that is protected from wind where the leaves tend to remain wet longer. Any area where it is difficult to apply fungicides such as edges and corners or under power lines if using aerial application should be examined. Scouts should look for large, black or purplish lesions on stems or leaves and the telltale cottony, white mold

growth, usually on the undersides. Be sure to check leaves and stems under the crop canopy as that is where the disease is most likely to begin.

- Use of a good protectant fungicide programme is necessary to fully protect any crop of potatoes or tomatoes.
- With potatoes, make sure that vines have been completely dead for 2-3 weeks prior to harvest. Fungicide applications should be continued until vines are dead. When foliage dies, spores of the late blight fungus that remain on the foliage also die. This practice will prevent infection of tubers during harvest and development of late blight in storage.

BACTERIAL SPOT, SPECK, AND CANKER OF TOMATOES

Bacterial spot, bacterial speck, and bacterial canker are widespread diseases of tomato that can cause localized epidemics during warm (spot and canker) or cool (speck), moist conditions. Bacterial spot can cause moderate to severe defoliation, blossom blight, and lesions on developing fruit. Bacterial speck also causes these symptoms but is usually not as severe as bacterial spot. Bacterial canker causes wilt, vascular discolouration, scorching of leaf margins, and lesions on fruit.

Symptoms

Foliar symptoms of bacterial spot and speck are identical. Small, water-soaked, greasy spots about 1/8 inch in diameter appear on infected leaflets. After a few days, these lesions are often surrounded by yellow halos and the centres dry out and frequently tear. Lesions may coalesce to form large, irregular dead spots.

In mature plants, leaflet infection is most concentrated on fully-expanded and older leaves and some defoliation may occur. Spots may also appear on seedling stems and fruit pedicels. In some cases, blossom blight may occur, causing flower abortion. This is more severe with bacterial spot and may result in a split fruit set which is especially troublesome with determinate cultivars intended for mechanical harvest. Bacterial spot and speck can usually be differentiated by symptoms on immature fruits.

Bacterial spot lesions are small water-soaked spots that become slightly raised and enlarged until they are about 1/4 inch in diameter. The centres of these spots later become irregular, light brown, slightly sunken with a rough, scabby surface. In the early stages of infection, a white halo may surround each lesion at which time it resembles the fruit spot of bacterial canker. Small lesions which have not yet become scabby are often confused with lesions of bacterial speck. Bacterial speck appears on immature fruit as a black, slightly sunken stippling, eventually causing lesions less than 1/16 inch in diameter. Fruit lesions are not initiated on mature fruit in either disease. Primary or systemic symptoms of bacterial canker (from infections originating in seeds or young seedlings) include stunting, wilting, vascular discolouration, development of open stem cankers, and fruit lesions. When affected stems are split open lengthwise, a thin, reddish-brown discolouration of the vascular tissue is observed, especially at the base of the plant. On young seedlings in the greenhouse, lesions may appear as raised pustules on leaves and stems.

These plants rarely survive the season in the field. Secondary symptoms in the field include leaf "firing" (necrotic marginal leaf tissue adjacent to a thin band of chlorotic tissue) and fruit lesions. Spots on fruit are relatively small (1/32 to 1/16 inch) surrounded by a white halo ("bird's-eye" spots). Canker bacteria may also invade internal fruit tissues, causing a yellow to brown breakdown.

Causal Organisms

Bacterial spot is caused by the bacterium, Xanthomonas campestris pv. vesicatoria, which can be carried as a contaminant on the surface of infested seed and has been found to overwinter in soil associated with plant debris. Bacterial speck is caused by another bacterium, Pseudomonas syringae pv. tomato.

This bacterium may also be seedborne and can overwinter on plant debris in soil and on the roots of many perennial plants. Bacterial canker is caused by Clavibacter michiganensis subsp. michiganensis, which, unlike the spot and speck pathogens, has the ability to infect tomato plants systemically. It is seedborne and can survive on infested plant debris in soil.

All three organisms may exist at low populations on leaf surfaces of symptomless plants. At the onset of favourable conditions, these low populations can increase rapidly and bacteria can then enter plants through stomata or small wounds and begin infection. Bacteria can spread rapidly with spattering rain and widespread epidemics may develop.

Penetration of tomato fruit occurs through wounds created by windblown sand, breaking of hairs, or by insect punctures. Optimal conditions for bacterial spot and canker are high moisture, high relative humidity and warm temperatures (75 to 90 degrees F). Bacterial speck is more likely to occur under cool (64 to 75 degrees F), moist conditions.

- Rotate tomatoes with non-solanaceous crops with at least 2 to 3 years between tomato crops. Avoid rotation with peppers, which are also susceptible to bacterial spot.
- Plant only seed from disease-free plants or seed treated to reduce any bacterial populations. Treatments include:
  - Fermentation of tomato pulp and seeds at room temperature for 4–5 days;
  - Soaking seeds in 0.6–0.8 per cent acetic acid for 24 hr at 70 degrees F;

  - Soaking seeds 5ñ10 hr in 5 per cent hydrochloric acid;
  - Hot water treatment of seeds (122 degrees F for 25 minutes); or
  - Sodium hypochlorite (bleach) treatment [20–40 minute soak of seeds in 1 per cent sodium hypochlorite (20 per cent bleach)]. Some decrease in germination may be expected from these treatments.
- Use only transplants free of disease symptoms.
- Carry out proper sanitation of transplant production greenhouses. Remove all weeds and plant debris, clean all tools with disinfectant solution, and wash hands thoroughly before and after handling plants. Water plants early in the day to reduce the amount of time foliage is wet.

  Do not handle plants when they are wet. After each crop, clean greenhouse walls, benches, etc., with hot soapy water, followed by thorough rinsing and treatment with a disinfectant. If possible, close up greenhouse after transplant production is completed to allow natural heating during the summer. Use only new plug trays and pathogen-free planting mixes. Avoid growing peppers and tomatoes in the same greenhouse unless pepper seed has also been treated as in step 2.
- In the field, control irrigation to minimize the time foliage is wet and avoid working among wet plants to minimize spread of disease.
- Applications of mancozeb plus copper soon after transplanting may help retard development and spread of bacterial spot and speck. This practice is not particularly effective for management of bacterial canker. Many tomato processors will not accept tomatoes treated with mancozeb or other EBDC fungicides. Check with your processor before applying one of these fungicides.

ANTHRACNOSE OF TOMATO

Anthracnose is a common and serious disease of tomato fruit. This disease can occasionally cause severe damage to peppers, especially when red fruit is allowed to develop. Anthracnose can reduce a bountiful harvest into rotted fruit in a few days in warm, moist weather.

Symptoms

Small, watersoaked, circular lesions develop under the skin of fruit as it ripens. These become sunken and dark. Numerous dark specks, the fruiting bodies of the fungus, develop in the lesions in concentric rings. In moist, warm weather, these black bodies ooze gelatinous pink spore masses. In warm weather the fungus and soft rot bacteria which enter the split skin over the lesions spread internally forming a semisoft decay which renders the fruit worthless.

Causal Organism

Colletotrichum coccodes survives between crops on infested plant debris in the soil. Early in the growing season, spores from the soil splash on lower leaves of the tomato plant. Few symptoms develop on infected leaves, but the spores produced on foliage can be carried by splashing rain to developing green fruit. Infected green fruit will not develop symptoms of anthracnose until they begin to ripen. Ripe fruit is very susceptible to this fungus.

Control

- Rotate 3 years between pepper and tomato crops.
- Plant tomatoes and peppers in well drained fields to avoid excess soil moisture as fruit ripen.
- Apply overhead irrigation during the early part of the day so that plants dry before sundown.
- Harvest and use fruit before it fully ripens.
- If conditions favour development of anthracnose, a preventative spray programme may be required to give adequate control of this disease. Apply registered fungicides according to product label instructions when weather conditions are above 65 degrees F and the foliage is likely to remain wet longer than 6 hours.
  Applications to tomatoes should begin when the first fruit is larger than a walnut. Applications to peppers should be started as soon as fruit is present.

BLOSSOM-END ROT OF TOMATO, PEPPER, AND EGGPLANT

Blossom-end rot is a serious disorder of tomato, pepper, and eggplant. Growers often are distressed to notice that a dry sunken decay has developed on the blossom end of many fruit, especially the first fruit of the season. This nonparasitic disorder can be very damaging, with losses of 50 per cent or more in some years.

Symptoms

On tomato and eggplant, blossom-end rot usually begins as a small water-soaked area at the blossom end of the fruit. This may appear while the fruit is green or during ripening. As the lesion develops, it enlarges, becomes sunken and turns black and leathery. In severe cases, it may completely cover the lower half of the fruit, becoming flat or concave.

Secondary pathogens commonly invade the lesion, often resulting in complete destruction of the infected fruit. On peppers, the affected area appears tan, and is sometimes mistaken for sunscald, which is white. Secondary molds often colonize the affected area, resulting in a dark brown or black appearance. Blossom end rot also occurs on the sides of the pepper fruit near the blossom end.

Cause

Blossom-end rot is not caused by a parasitic organism but is a physiologic disorder associated with a low concentration of calcium in the fruit. Calcium is required in relatively large concentrations for normal cell growth. When a rapidly growing fruit is deprived of necessary calcium, the tissues break down, leaving the characteristic dry, sunken lesion at the blossom end.

Blossom-end rot is induced when demand for calcium exceeds supply. This may result from low calcium levels or high amounts of competitive cations in the soil, drought stress, or excessive soil moisture fluctuations which reduce uptake and movement of calcium into the plant, or rapid, vegetative growth due to excessive nitrogen fertilization.

- Maintain the soil pH around 6.5. Liming will supply calcium and will increase the ratio of calcium ions to other competitive ions in the soil.
- Use nitrate nitrogen as the fertilizer nitrogen source. Ammoniacal nitrogen may increase blossom-end rot as excess ammonium ions reduce calcium uptake. Avoid over-fertilization as side dressings during early fruiting, especially with ammoniacal forms of nitrogen.
- Avoid drought stress and wide fluctuations in soil moisture by using mulches and/or irrigation. Plants generally need about one inch of moisture per week from rain or irrigation for proper growth and development.
- Foliar applications of calcium, which are often advocated, are of little value because of poor absorption and movement to fruit where it is needed.

BACTERIAL RING ROT OF POTATOES

Bacterial ring rot is an important disease of potatoes and is one of the main reasons for rejection of seed potatoes from certification programmes. This disease is particularly serious because it has the potential to spread quickly throughout a farm and may lead to severe losses if left unchecked. Ring rot was originally found in Germany in the late 1800's. The causal bacteria were introduced into the United States in the early 1930's and by 1940 were found throughout the country.

SYMPTOMS

Severe ring rot can result in wilting of leaves and stems along with yellowing and death of leaves. Lower leaves usually wilt first, are slightly rolled at the margins, and are paler green than healthy leaves. As wilting progresses, leaf tissues between veins become yellow. In the later stages of disease, margins of lower leaves die and become brittle, and eventually entire stems yellow and die. Frequently, only one or two stems in a hill will develop symptoms and, in some cases, there are no above-ground symptoms at all. Ring rot derives its name from a characteristic breakdown of the vascular ring within the tuber. This often appears as a creamy-yellow to light-brown, cheesy rot. The symptom is most frequently observed when a diseased tuber is cut crosswise at the stem end.

In severe cases, the vascular ring may be separated, and a creamy or cheesy exudate can be forced out from this tissue when the tuber is squeezed. On the outer surface, severely diseased tubers may show slightly sunken, dry, cracked areas. Infected tubers are often invaded by secondary decay organisms which may lead to complete breakdown. Symptoms of ring rot in the vascular tissue of infected tubers are often less obvious t, appearing as only a broken, sporadically appearing dark line, or as a continuous, yellowish discolouration. Because of this, laboratory tests should always be performed to confirm a diagnosis of ring rot.

Causal Organism

Ring rot is caused by the bacterium Clavibacter michiganense subsp. sepedonicus. Ring-rot bacteria survive between seasons mainly in infected seed tubers. They are also capable of surviving 2-5 years in dried slime on surfaces of crates, bins, burlap sacks, or harvesting and grading machinery, even if exposed to temperatures well below freezing. Survival is longest under cool, dry conditions. Ring-rot bacteria do not survive in soil in the absence of potato debris, but can survive from season to season in volunteer potato plants. Wounds are necessary for penetration of the bacteria into seed pieces.

The pathogen is easily transmitted from diseased tubers to healthy seed pieces during the seed-cutting process. A knife that cuts one infected tuber can spread these bacteria to the next 20-100 seed pieces. Likewise, the bacteria may be spread during planting, particularly if a picker-type planter is used. Ring-rot bacteria can be moved in irrigation water and by chewing insects, such as Colorado potato beetles and flea beetles. After the bacteria become established in a plant, they multiply and move throughout the plant via the water-conducting tissues. Fortunately ring-rot bacteria are capable of causing disease only in potato, although they may be able to colonize roots of sugar beets.

- Plant only certified disease-free seed tubers. Certified seed potatoes are produced under regulations mandating zero tolerance for ring rot. Although use of certified seed tubers will not guarantee total freedom from ring rot bacteria, it is the best assurance.
- Discontinue use of any lot of seed tubers in which ring rot is found. Seed lots known to be contaminated with ring-rot bacteria should never be planted.
- Before handling seed tubers, all containers, tools, knives and mechanical cutters, planters, and other equipment should be thoroughly washed with a detergent solution, rinsed, and then sanitized with a disinfectant (for current recommendations,
- If ring rot is confirmed to be present, a thorough cleanup must be undertaken. Dispose of all infected tubers away from potato

production areas. Clean all surfaces of storages and equipment to remove all mud, dirt and debris and then wash with a strong detergent in hot water applied by a high-pressure washer. After cleaning, sanitize all storages and equipment with a disinfectant. Do not plant potatoes for two seasons in any field in which ring rot has been found.

POTATO PINK ROT, PYTHIUM LEAK AND SEED-PIECE DECAY

Pink rot and Pythium leak, sometimes collectively called water rot, occur sporadically wherever potatoes are grown. These diseases are a problem of mature tubers at harvest and in storage. They are most serious when warm, wet soil conditions persist during tuber formation and at harvest. When newly-planted seed pieces are exposed to these conditions, Pythium seed-piece decay also can be severe. Major problems with these diseases are usually associated with excessive rainfall or irrigation either early or late in the season, especially on poorly-drained soils.

SYMPTOMS

Pythium seed-piece decay often results in delayed emergence and poor stands. Infected seed pieces become a soft, watery mass in the soil. Symptoms of pink rot in mature plants include brown or blackened roots or stolons, and in severe cases, leaf chlorosis, stunting, wilting, and even plant death. Tubers develop pink rot mostly through diseased stolons, but occasionally infections occur at buds or lenticels. Decay spreads through infected tubers with the advancing margin of the rot usually sharply defined by a dark line, which may be visible through the skin.

Eyes of infected tubers are often dark brown. Decaying tubers remain intact, but are spongy and odorless. If squeezed, a clear liquid will exude. When infected tubers are cut open, the internal tissues turn salmon pink after a 15-20 minute exposure to air, then later become brownish-black. Pythium leak usually develops through harvest wounds in tuber surfaces and begins as a discoloured, watersoaked area.

As with pink rot, the advancing margin of infection is usually bounded by a dark line. Infected tissues are extremely watery, and appear brown or gray. Severely rotted tubers are of a uniform texture resembling a soft, watery paste.

Causal Organisms

Pink rot is caused by several species of the soilborne fungus Phytophthora (NOT the species that causes late blight) while Pythium leak and seed-piece decay are caused by several species of the closely related soilborne fungus Pythium. These fungi are widely distributed in both water and soil, and their behaviour is similar. They survive in soil within decaying plant material, or as resistant spores free in the soil. In warm, moist soil, these fungi produce swimming spores that move in water films. Roots can be infected by Phytophthora at almost any stage of plant growth, but symptoms are more severe on younger roots.

Both groups of fungi infect tubers through wounds, but Phytophthora generally infects tubers before harvest, often through stolons. Infection by Pythium usually occurs through harvest wounds, especially at temperatures above 70 F. Seed pieces can be infected by Pythium as soon as they are planted.

- Select areas with well-drained soils for planting potatoes.
- Use a crop rotation away from potatoes for at least 4 years if pink rot or leak have been severe. This may reduce the amount of fungus surviving in the soil.
- Delay planting for at least 2 weeks after plowing down green vegetation as this may temporarily stimulate populations of Pythium fungi.
- Avoid planting in soils colder than 45 F or warmer than 70 F.
- Avoid harvesting infested fields when soils are especially wet or soil temperatures are below 50 F or above 65 F. Stop irrigation well in advance of harvest.
- Avoid bruising tubers during harvest by adjusting equipment properly, keeping digger chains fully loaded and minimizing drops to 6 inches or less. Do not leave harvested tubers lying on warm, moist soils for any length of time as infection with Pythium may occur quickly.
- Leave low spots in fields unharvested if they have been waterlogged and much rot is present.
- Keep tubers cool and as dry as possible during harvest, loading, transit and storage.
- Grade out infected tubers as much as possible before placing harvested tubers in storage.

  Store lots of harvested tubers containing many infected tubers separately from healthy lots. Good airflow through the pile should be provided to dry out leaky tubers. Lots with significant amounts of disease should be marketed as soon as possible as they will not store well.

SCAB OF POTATO TUBERS

Scab is a disease of potato tubers that results in lowered tuber quality due to scab-like surface lesions. There are no above-ground symptoms. Two forms of scab occur.

Common scab occurs in all production areas and is most severe in soils with a pH above 5.5. Another less common form, called acid scab, is important in acidic soils.

SYMPTOMS

Scab symptoms are quite variable. Usually, roughly circular, raised, tan to brown, corky lesions of varying size develop randomly across tuber surfaces. Sometimes scab develops as a rather superficial layer of corky tissues covering large areas of the tuber surface. This is called russet scab. Pitted scab can also occur where lesions develop up to 1/2 inch deep. These deep lesions are dark brown to black, and the tissues underneath are often straw-coloured and somewhat translucent. More than one of these lesion types may be present on a single tuber.

Although scab symptoms are usually noticed late in the growing season or at harvest, tubers are susceptible to infection as soon as they are formed. Small brown, water-soaked, circular lesions are visible on tubers within a few weeks after infection. Mature tubers with a well-developed skin are no longer susceptible, but existing lesions will continue to expand as tubers enlarge. Thus disease severity increases throughout the growing season. Scab is most severe when tubers develop under warm, dry soil conditions. Coarse-textured soils that dry out quickly are therefore more conducive to scab than are fine-textured soils.

A few other conditions can be confused with scab. White, enlarged lenticles, which frequently occur on potato tubers harvested from wet soil, can be mistaken for scab. Usually this condition will disappear when tubers are dried. Patchy russeting, checking, or cracking of tuber surfaces caused by the fungus Rhizoctonia also may be confused with russet scab. A totally different but uncommon disease called powdery scab, caused by the fungus Spongospora subterranea, causes very similar scab-like symptoms. Laboratory examination may be necessary to identify these diseases.

Causal Organisms

Scab is caused by a group of filamentous bacteria called actinomycetes that occur commonly in soil. In soils with a pH above 5.5, Streptomyces scabies is usually responsible for common scab, and is capable of causing all the types of scab lesions. It is commonly introduced into fields on seed potatoes, and will survive indefinitely on decaying plant debris once the soil is contaminated. Because the organism can survive passage through the digestive tract of animals and be distributed.

Blackleg, Aerial Stem Rot, and Tuber Soft Rot of Potato

Blackleg, aerial stem rot, and tuber soft rot are all similar diseases caused by several types of soft-rot bacteria. Blackleg and tuber soft rot occur wherever potatoes are grown. Aerial stem rot is also widespread, but is most severe under sprinkler-irrigation.

Symptoms

Blackleg begins from a contaminated seed piece, but the symptoms can occur at several stages of plant development. In severe cases, entire seed pieces and developing sprouts may rot in the ground prior to emergence, resulting in a poor stand. Blackleg often develops after plants are well up or even in flower.

In this case, stem bases of diseased plants typically show an inky-black to light-brown decay that originates from the seed piece and can extend up the stem from less than an inch to more than two feet. Leaves of infected plants tend to roll upward at the margins, become yellow, wilt, and often die. Aerial stem rot (also called bacterial stem rot or aerial blackleg) is initiated by soft-rot bacteria from sources external to the seed piece. Stem infection can occur through wounds or through natural openings such as leaf scars. Lesions on diseased stems first appear as irregular brownish to inky-black areas.

These enlarge into a soft, mushy rot that causes entire stems to wilt and die. Potato tubers with soft rot have tissues that are very soft and watery, and have a slightly granular consistency. The diseased tissue is cream- to tan-coloured, and often has a black border separating diseased from healthy areas.

In the early stages, soft-rot decay is generally odorless, but later a foul odor and a stringy or slimy decay usually develops as secondary decay bacteria invade infected tissues. Most internal tuber tissues may be consumed by soft rot organisms, sometimes leaving only a shell of skin remaining in the soil.

Causal Organisms

Blackleg, aerial stem rot, and tuber soft rot are caused by two closely related bacteria, Erwinia carotovora subsp. atroseptica and Erwinia carotovora subsp. carotovora. E. c. carotovora is very common and has an extensive host range, including most fleshy vegetables. It survives readily in soil and surface waters such as rivers, lakes, and even oceans. These bacteria are capable of multiplying and persisting in the root zones of many host and nonhost crop and weed species. In contrast, E. c. atroseptica is associated mostly with potatoes.

These bacteria do not survive well in soil for more than one year, unless they are contained within diseased tubers or other potato plant debris. Blackleg is usually caused by E. c. atroseptica carried on contaminated seed tubers. Most lots of seed tubers are contaminated to some degree, but the bacteria are usually dormant and do not cause disease unless environmental conditions are favourable. In contrast, aerial stem rot is usually caused by E. c. carotovora contained in infested soil or introduced to the crop by irrigation water, wind-blown rain, and insects.

Tuber soft rot can be caused by either of these soft-rot bacteria.

Moisture and temperature are the two critical factors in initiation and development of soft-rot diseases. High soil temperatures and bruising of seed tubers favour seed-piece decay and pre-emergence blackleg. Blackleg in growing plants is favoured by cool, wet soils at planting followed by high temperatures after emergence.

Dense plant canopies and long periods of leaf wetness favour infection of aerial plant parts. Although tuber soft rot can occur at any temperature above 50 F, disease develops best above 75 F. Oxygen depletion in tubers also favours soft rot. When seed pieces in soil or tubers in storage become covered with a film of water, the tissues rapidly become depleted of oxygen.

This also may be induced by soil flooding or improper drying of washed tubers. Once it starts, tuber soft rot can proceed rapidly in storage. "Wet" areas may develop in the piled tubers that flow onto ones below, spreading the bacteria. Heat, coupled with condensation on tuber surfaces, can further adversely affect storage conditions, resulting in accelerated "melt" of the pile.

- Plant only certified, disease-free seed tubers. If possible, use whole (B-size) seed tubers that do not have to be cut.
- When receiving seed tubers in bags, do not stack more than five bags high. With bulk or bagged seed, store at 40-45 F until 2-3 weeks before planting, then warm to 55-60 prior to cutting.
- Clean all equipment used for cutting seed tubers thoroughly and then sanitize with an appropriate disinfectant.
- Treat cut seed pieces with recommended fungicide dressings immediately after cutting.
- Plant treated cut seed pieces immediately if soil temperatures are 55-65 F at planting depth. Seed pieces can be held 1-2 weeks at 55-60 F and 95-99 per cent relative humidity to hasten healing of cut surfaces. Condensation on surfaces of seed pieces must be avoided.
- Do not irrigate fields until plants are well emerged. Avoid using surface water for irrigation.
- During crop growth, monitor irrigation and nitrogen fertility to minimize excessive vine growth that will promote leaf wetness within the plant canopy.
- Harvest tubers only after the vines are completely dead to ensure skin maturity. Low spots in the field should be left unharvested if significant waterlogging has occurred.
- Take all precautions to minimize cuts and bruises when harvesting and handling tubers.
- Hold newly harvested potatoes at 55-60 F with 90-95 per cent relative humidity for the first 1-2 weeks to promote wound healing. After this curing period, lower the temperature of table stock to 38-40 F for long-term storage. Never wash tubers prior to storage.

SEED-PIECE DECAY OF POTATO

Fusarium dry rot is an important postharvest disease of potato tubers that causes significant losses in storage and transit of both seed tubers and those for table consumption. It is also a major cause of seed-piece decay after planting.

SYMPTOMS

Infected tubers usually develop a dry rot, but a moist rot may occur if secondary infections with soft-rot bacteria also are involved. Surfaces of infected tubers are sunken or wrinkled, and rotted tissues appear brown or gray to black.

A white or pink mold is sometimes visible on tuber surfaces. When tubers are cut, internal cavities within rotted tissues may contain white, yellow or pink molds. In storage, blue, black, purple, gray, white, yellow, or pink spore masses may develop in these internal cavities. After low-temperature storage, internal tissues often will become firm and dry or even powdery.

Causal Organisms

Fusarium dry rot is caused by several species of the soilborne fungus Fusarium. These fungi are common in most soils where potatoes are grown and survive as resistant spores free in the soil or within decayed plant tissues. Although some infections may develop on tubers before harvest, most infections occur as the fungus enters tubers through harvest wounds.

Small, brown lesions appear at wound sites 3-4 weeks after harvest and continue to enlarge during storage, taking several months to develop fully. The disease develops fairly rapidly at temperatures above 50 F, but lesions will cease enlarging below 40 F. The fungus is only dormant at these low temperatures, however, and will resume growth when tubers are warmed. Fusarium seed-piece decay is really the same disease. Seed tubers may be infected prior to shipment. Decay during transit or storage often accounts for poor quality seed tubers. When these Fusarium fungi are present on seed pieces or in the soil, poor stands may result, especially if cut surfaces of seed pieces are not properly healed.

Fusarium seed-piece decay begins as reddish-brown to black depressions on cut surfaces. These may expand to cover the entire seed piece and often result in a slimy rot when infection by secondary soft-rot bacteria follows.

- Harvest tubers only after the vines are completely dead to ensure skin maturity.
- Take all precautions when harvesting and handling tubers to minimize cuts and bruises.
- Hold newly harvested potatoes at 55-60 F with 90-95 per cent relative humidity for the first 1-2 weeks to promote wound healing. After this curing period, lower the temperature of table stock to 38-40 F for long-term storage.
- Plant only certified, disease-free seed tubers. If possible, use whole (B-size) seed tubers that do not have to be cut into seed pieces before planting.
- When receiving seed tubers in bags, do not stack more than five bags high. With bulk or bagged seed, store at 40-45 F until 2-3 weeks

before planting. Then allow seed potatoes to warm prior to cutting.

- Treat cut seed pieces with recommended fungicide dressings immediately after cutting.
- Plant treated cut seed pieces immediately or store them at 55-60 F and 95-99 per cent relative humidity to hasten healing of cut surfaces. Condensation on surfaces of seed pieces must be avoided.

RHIZOCTONIA STEM AND STOLON CANKER OF POTATO

Rhizoctonia stem and stolon canker, also called black scurf, is an extremely common problem on potatoes that can result in delayed emergence, reduced stands and poor tuber quality.

Symptoms

Rhizoctonia canker often goes unnoticed until harvest when tubers are found to be covered with small brownish-black fungal bodies (sclerotia) that look somewhat like bits of black soil that will not wash off. These may vary in size from a pinhead to as large as a pea. When introduced into soil, these structures can survive and germinate the following spring to attack young shoots, roots, stolons, and tubers of the new crop. In many cases, weak, spindly-looking or late-emerging plants may be the result of an attack by this fungus.

The first sprouts are often killed before they reach the surface, resulting in the emergence of a weaker secondary sprout. Dry, sunken, brownish lesions developing on the base of the stem below the soil line are evidence of the stem canker phase of this disease. The lesions may girdle the stem or large cankers may interfere with movement of nutrients from the leaves to the tubers. In such cases, vines become yellow to reddish purple; the leaves begin to curl upward; the stalks swell, particularly at the nodes; and often small purplish tubers form where leaves branch from stems.

During midseason under a dense canopy of foliage, the fungus may develop a white, powdery mold growth on stems, extending just above the soil line. Cankers that form on stolons may prune off young developing tubers. This disease may lead to russeting or surface cracking of mature tubers and sometimes shallow, brown lesions will form around lenticles. Low soil temperatures in the fall favour formation of sclerotia on tubers.

Causal Organism

This disease is caused by the fungus Rhizoctonia solani which can survive in the soil for many years, even under relatively dry conditions. Rhizoctonia causes disease in a wide variety of crops, but the strains found in association with potato generally do not attack and reproduce on other plant species. The fungus survives in soil associated with decomposing plant residues. In addition, the sclerotia can survive on infected tubers and can persist free in soil for extended periods. Emerging sprouts are usually attacked by fungus present on the seed tubers. Once green leaves develop on sprouts, stem tissues are much less susceptible to infection. Stem cankers, stolon infections, and sclerotia on tubers usually develop when these tissues grow in proximity to sources of the Rhizoctonia fungus in soil. Cool (55-60 F), moist soils are optimal for infection. Sclerotia form on the surfaces of mature tubers under cool, moist conditions, generally after the vines have begun to die.

- Use a crop rotation with corn, grasses, and cereal grains. If this disease has been severe, 3-5 years should elapse between potato crops.
- Plant certified seed tubers that are free of Rhizoctonia on the skin.
- Use planting practices that promote rapid emergence: Avoid planting in heavy, poorly-drained soils. Plant seed potatoes when soil is warm (above 60 F). Cover seed tubers with no more than 2 inches of soil.
- Harvest tubers promptly after vines are dead to avoid the development of sclerotia on the surfaces of tubers while still in the soil.

MOSAIC VIRUS DISEASES OF VINE CROPS

Mosaic diseases of vine crops are caused by at least five different viruses:

- Cucumber mosaic virus (CMV),
- Squash mosaic virus (SQMV),
- Watermelon mosaic virus-2 (WMV-2),
- Zucchini yellow mosaic virus (ZYMV)
- Papaya ringspot virus-watermelon isolate (PRSV-W). All five of these viruses can infect all cultivated vine crops (squash, melons, gourds, cucumbers, pumpkins) and under ideal conditions can cause a high rate of crop failure and severe economic losses.

Symptoms

As the names imply, all five viruses cause a mottling of foliage called mosaic. This is characterized by the presence of intermingled patches of normal and light green or yellowish coloured plant tissue.

Depending on environmental conditions, mosaic symptoms can range from mild to severe and be visible on both leaves and fruit. The younger the plant when infected, the more severe the symptoms as the plant matures. In some cases, plants infected at the seedling stage may collapse and die. Plants infected at the flowering stage may not set fruit or young fruits may abort.

If plants are more mature when infected they do not show severe mosaic and may still produce marketable fruit. The most dramatic symptoms are often associated with infected fruit. Fruit symptoms can range from subtle colour change to severe deformation. It is not uncommon to have two or more viruses infecting the same plant and in these cases symptoms may be much more severe than if the plant were infected with only one virus.

It is nearly impossible to distinguish between any of the five viruses based only on visible symptoms. Because of this, it is important to have suspect plants sent to a diagnostic clinic capable of using specialized techniques to identify the viruses that are present. Proper disease management depends on knowing which specific virus is involved.

Disease Cycle and Transmission

All five vine crop mosaic viruses are transmitted by insects. CMV, ZYMV, WMV-2 and PRSV-W are transmitted primarily by the green peach aphid and the melon aphid. SQMV is transmitted by the striped cucumber beetle and spotted cucumber beetle and also through seed, but only at a very low rate. In most cases these insects acquire the virus by feeding on weeds or other vine crops that are infected.

The insects then feed on noninfected vine crops and in the process of this feeding transmit the virus to the host plant. In the case of the green peach and melon aphid, the insect carries the virus on its probing mouthparts and the aphid only needs to probe the plant to transmit the virus.

Feeding by the aphid is not necessary. Once the insect probes one plant, it can move onto the next, probing and infecting as it goes. The more aphids present, the quicker and more complete the virus spread. By this process, large acreages can become infected quickly.

- Plant varieties with resistance to these viruses whenever possible. The availability of varieties with virus resistance varies depending on the virus and the vine crop in question. Major seed producers list resistant varieties in their seed catalogs.
- Control weeds which could harbor the viruses from areas around production fields. Pokeweed is a primary weed in the area which harbors many plant viruses. Weeds also serve as hosts for insects that transmit the viruses.
- Since the virus is moved about by insects it would seem logical that insect control would be a primary method of virus control. However, in this case it is not that simple. Since aphid vectors need only probe the plant to transmit the virus, insect control is not really that effective in controlling the virus. Even if the aphid were to die immediately following probing, virus transmission would still take place. To control the insect effectively insecticides would need to be applied on a daily basis. The economics of this may not be feasible. Aphids also move long distances with weather fronts making local control difficult.

DAMPING-OFF AND ROOT ROT OF BEANS

Root rots, damping-off before and after seedling emergence, and seed rots are destructive diseases of green, snap, lima, and dry beans. These diseases are caused primarily by soilborne fungi.

Significant losses may occur to susceptible varieties, especially if cool, wet weather conditions prevail for the first few weeks after seeding and then are followed by hot, dry weather.

Disease incidence and severity often vary greatly, even in areas with a history of root rot. In the same growing season, it is not uncommon to lose a crop completely and then re-seed and experience no problems.

This situation results from changes in biological, environmental, and soil conditions. Since there are no commercially acceptable resistant varieties, growers should learn how to recognize these diseases and use a combination of management practices to minimize potential losses.

Symptoms

Damping-off before emergence results from fungal attack of germinating seeds and/or young seedlings while they are still in the ground. Infected seeds may fail to germinate, become soft and mushy, and finally disintegrate. Slightly darkened water-soaked lesions may be visible on stems of young seedlings. Infected areas enlarge quite rapidly, and seedlings may die shortly after infection, prior to emergence from soil. Roots or stems of seedlings that have already emerged also can be attacked at or below the soil line resulting in damping-off.

Infected roots are usually discoloured or rotted and sometimes reddish-brown lesions develop on the tap root.

Infected stem tissues are soft and colourless to dark-brown. Basal portions of invaded stems may be much thinner than the areas above the lesion, a condition called "wire stem."

As a result, the seedling may fall over and die. Damping-off is a major cause of poor stand establishment in bean plantings. Older plants can also be attacked by these fungi.

Later infections are usually confined to roots, which may result in stunting, wilting, or plant death. To diagnose bean root rots, suspected plants should be carefully dug and washed, because pulling plants may leave tissues with characteristic symptoms in the soil. If plants are brought to a diagnostic clinic, they should be dug and left intact in soil.

Causal Organisms

Root rots, damping-off, and seed rots are caused primarily by soilborne fungi. Fusarium root rot is caused by Fusarium solani f. sp. phaseoli. This fungus is capable of surviving long periods in soil, even in the absence of beans, by the production of thick-walled resting spores. Its host is mainly green beans, but lima beans and garden peas are also susceptible.

Coarse-textured, acidic and poorly fertilized soils favour development of Fusarium root rot. A number of fungi in the genus Pythium are capable of inciting seed rot and seedling damping-off. This may vary with environmental conditions and developmental stage of the host.

These fungi can survive in soil for many years, either by producing thick-walled resting spores, or in a vegetative condition within crop residues left from previous years. They may also survive by attacking several other garden vegetable crops such as beets, cabbage, peas, melons, squash, and cucumbers.

Development of diseases caused by Pythium is more prominent under:

- High or low temperatures that are unfavourable to crop growth,
- Excess water and/or nitrogen,
- Continued cropping of susceptible plants.

Rhizoctonia solani is another soil-inhabiting fungus causing root rots and damping-off. This fungus overwinters free in the soil or within decayed plant tissues. Aside from bean, its host range includes beets, cabbage, lettuce, peas, pepper, tomatoes, and many others. Warm, relatively dry soil conditions favour disease development.

- Do not grow beans or other susceptible crops continually in the same location. Continuous cropping of susceptible plants will eventually lead to a buildup of these fungi in the soil. Since they are capable of long-term survival, a rotation of 4-5 years is desirable. Avoid planting beans in fields known to be heavily infested with bean root-rot fungi.
- Plant beans only on well-drained soils or try to improve drainage. This could be done by improving soil structure and/or installing drain tiles. Subsoiling to a depth below the plowed layer will reduce soil compaction, and improve drainage. Deep plowing of the previous years' crop residues will reduce bean root rot.
- Delay planting until the soil is warm (above 65 F) and seed shallow to insure rapid emergence. Avoid planting seeds too close together-follow instructions on the seed container. Do not overfertilize, especially with nitrogen.
- Use of fungicide-treated seeds will minimize problems with damping off and root rots.

POWDERY MILDEW OF VINE CROPS

Powdery mildew can be a serious problem on foliage and stems of cucumbers, melons, squash, pumpkins, and ornamental gourds.

Symptoms

Small patches of fine, white threads develop on surfaces of infected leaf blades in late July on most vine crops. These patches grow together and eventually cover stems and foliage with white, powdery masses of spores. In severe infections, leaves will yellow and die. Petioles, stems and, rarely, fruit will also become infected. Early death of leaves can decrease the total amount of fruits formed as well as reduce fruit size. Loss of foliage hastens maturity of fruit and increases sunburning. Stems infected with Powdery Mildew become brown. This can spoil the appearance of the "handles" on pumpkins which lowers value of the fruit.

Causal Organisms

Two fungi cause powdery mildew on vine crops, Sphaerotheca fuliginea and Erysiphe cichoracearum. S. fuliginea is most common. Powdery mildew fungi thrive under conditions of high relative humidity, warm temperatures, low light, high fertility, and succulent plant growth.

Unlike bacteria and many fungi, free moisture on leaf surfaces actually inhibits infection by these pathogens, although very high relative humidity is required for spore germination. Powdery mildew fungi grow only on living host plants.

They survive the winter as dormant mycelium on perennial plants or as spores in thick-walled fruiting structures. Powdery Mildew spores can be blown into the spring from warmer southern areas. Mature foliage is most readily infected; very young leaves are nearly immune.

Control

1. Plant resistant cultivars of cucumbers and muskmelons. Pumpkin varieties vary greatly in susceptibility to this disease.
2. Use cultural practices that avoid excessive succulence, overcrowding, shading, overwatering, or excess fertilization especially with nitrogen.
3. Avoid making new planting of vine crops in the vicinity of older plantings, especially if mildew is already present.
4. Beginning in late July or early August when Powdery Mildew first appears, regular applications of fungicides may be required.

SEPTORIA LEAF SPOT OF TOMATOES

Septoria leaf spot is a disease of the foliage and stems. It does not affect fruit directly. The disease causes rapid defoliation when weather is warm and moist.

Symptoms

Although the symptoms may appear on the leaves and stems at any stage of plant development, they usually become evident after plants have begun to set fruit in mid-July. Initially the fungus causes small, water-soaked, roughly circular spots scattered randomly over the leaf. These spots enlarge to become 1/16 to 1/8 inch in diameter with dark margins and tan centres.

Small, dark, pimple-like fruiting bodies in which spores of the fungus are produced can be seen in the centre of the lesions. Older leaves near the ground usually show symptoms first. Symptoms appear rapidly on younger foliage in rainy weather. When leaves are heavily infected, they drop prematurely. Exposed fruits are subject to sunscald. The fungus which causes this disease can grow on weeds such as Jimson weed, horse nettle and nightshade.

Causal Organism

Septoria lycopersici lives between tomato crops in the soil on infested debris of tomato and weeds. Spores formed on crop debris splash onto foliage and start the disease. Wind and rain spread spores produced in the dark bodies formed in leaf spots to adjacent uninfected leaves. The fungus is most active between 60 and 80 degrees F when rainfall is abundant.

Control

- Rotate out of tomatoes for 4 years.
- Deep plow, preferably in the fall, to bury all plant refuse. Grow tomato transplants in sterilized soil.
- Control weeds, especially horse nettle, Jimson weed, and nightshade.
- Use of a protectant fungicide may be necessary to adequately control Septoria Leaf Spot when conditions are favourable for disease development.

FUSARIUM WILT OF VINE CROPS

Several different species of Fusarium, a soil-borne fungus, cause wilting of watermelon, muskmelon, cucumber, squash, and other vine crops. In many cases the fungus-causing wilt in a particular crop is specific to that crop. These fungi are generally capable of surviving for long periods in the soil.

Symptoms

Wilting of older plants is often the first symptom of the disease. Before diseased plants totally collapse, however, they may begin to wilt during the hottest part of the day and recover during the night. Infected plants are often stunted and yellowed. Leaves often have dead areas which can mimic nutrient deficiencies. Stems of wilted plants when cut lengthwise at the soil line may show brown discolouration in the woody tissues immediately under the bark. Vines killed by Fusarium can be covered with pinkish-white fungal growth in wet weather.

Causal Organism

Fusarium oxysporum is well adapted to life in the soil. It can survive season to season in old diseased vines or it can live free in infested soil for many years in the absence of its host crop. The fungus grows at soil moisture and temperature favourable for vine crop growth. If the soil is very wet, infection is reduced. The fungus is stimulated to germinate when roots of susceptible host plants are growing nearby. It enters the plant through root tips or where some opening is present.

Control

1. Plant wilt-resistant cultivars whenever possible. Degree of resistance is influenced by the populations of the fungi in the soil, and which races are present.

2. Avoid introduction of the fungus into fields. Once soil is infested, crop rotation will be of little use because of the long-term survival of these fungi in soil. The fungus can be spread on equipment, tools, feet, and surface water contaminated with infested soil. Do not put compost on fields which has been made from diseased vines. Compost resulting from such piles will contain the fungus.
3. Crop rotations of three to four years may be helpful in lowering the amount of Fusarium in the soil. Since the types of Fusarium-causing wilt in a crop are generally specific to that crop it is possible to rotate between vine crops in infested soil.
4. In commercial production of vine crops, soil fumigation may be useful.

ROOT KNOT NEMATODE

Root knot first attracted attention in 1855 when it was found in a greenhouse in England. Since then it has been a cause for concern over much of the world. It affects more than 2,000 species of plants including economically important forage crops, small grains, fruits, vegetables, field crops, nursery crops, and turf grasses. Root knot nematodes are cosmopolitan in distribution and occur widely in the United States.

Symptoms

Root knot is very distinctive because of the galls or swellings produced on roots and underground portions of stems. These deformations can often completely ruin crops for sale. Plants, if infected when young, will be stunted, more susceptible to drought stress, and show symptoms of nutrient deficiency. Large and small roots may be affected with swellings varying from round sphere-like galls to elongated spindles formed from large numbers of individual galls growing together. Root knot galls involve the entire root in the affected zone.

They do not take the form of easily detached galls like those produced by nitrogen-fixing bacteria on the roots of legumes. Root knot galls should not be confused with the fungal disease, club root, on cruciferous crops. In most instances, root knot is characterized by smaller swellings, and more uniformly distributed infection on the lateral feeding roots than is typically seen with clubroot.

When the galls formed by the root knot nematode are broken open, shiny white bodies about the size of a pinhead, the enlarged female nematodes, are usually found. Also, glistening white to yellow egg masses are present on the root surface. The galls formed by clubroot do not possess this characteristic and are usually pinkish or brick coloured.

Phenoxy-type herbicides such as 2,4 D can cause swellings on stems of cruciferous crops which superficially look like root knot galls. Herbicide damage on these crops will not affect the roots, however. On young potato tubers the outer surface appears rough and warty because of the enlarged females underneath the skin.

Causal Organism

Root knot is caused by a small round worm, Meloidogyne spp. There are about 40 species at the present time. Not all plants are susceptible to any one species. Juveniles emerging from eggs in the soil penetrate between and through cells to a position at the centre of the root usually near the growing tip. Feeding by the nematodes causes increases in root cell numbers and size. Enlarged cells, called giant cells, serve as food sources for female root knot nematodes.

During feeding, juveniles which will become females undergo a series of molts and enlarge. The female does not move from the feeding site. Each female deposits 300-500 eggs in a protective, jelly-like matrix at the root surface. Eggs, particularly in the egg mass, can withstand unfavourable environmental conditions. They represent one means of overwintering. Root knot nematodes may also overwinter in the soil in a juvenile stage. Root knot nematodes are often introduced into a growing area in infected planting stock.

Control

- Rotate at least 3 years with non-host crops. While the feeding habits of field populations must be determined before crop rotation plans are set, the most common root knot species will not reproduce on grasses, small grains, or corn. Clean cultivation to keep down weed hosts of the nematodes is important.
- Use transplants and nursery stock certified free of root knot nematode.
- Plant resistant varieties wherever possible.
- In commercial production, soil can be treated with fumigant-type nematicides.

PHYTOPHTHORA BLIGHT OF PEPPER AND CUCURBITS

Phytophthora blight, a highly destructive disease of peppers and cucurbits. It can become a serious problem during periods of heavy rainfall; the pathogen can spread rapidly through the crop, resulting in severe losses within a short time.

Symptoms

Peppers

Phytophthora blight affects both seedlings and mature plants. Infected seedlings show typical damping-off symptoms. Infection of older plants usually begins at or below the soil line. Water-soaked, dark brown lesions on the lower stems (collar rot phase) usually extend upward for an inch or more above the soil line and may expand to girdle the stems, preventing upward movement of water and nutrients.

This often results in a sudden wilting of foliage. Root infections may also occur which kill roots and cause wilting of the plant without the appearance of stem cankers. The foliar phase of this disease commonly occurs at forks in the branches, resulting in dark, girdling cankers and wilting of leaves and fruits. Infected leaves develop circular or irregular, dark green, water-soaked lesions which dry and appear light tan. Fruit lesions may also appear as enlarging, watersoaked areas, which then shrivel and darken.

A mass of white fungal growth may develop inside the fruit, and seeds usually turn dark brown or black. A fine, grayish-white to tan mold may also become evident over the lesion on the fruit surface. Under humid conditions, fungal growth develops extensively over the entire fruit.

Cucurbits

Seedlings show typical symptoms of damping-off. Mature plants may wilt suddenly as a result of root infection, in the absence of obvious stem or vine lesions. Stem and leaf petiole lesions are light to dark brown, watersoaked and irregular, eventually becoming dried and papery. Fruit symptoms begin as small, water-soaked circular lesions, which expand to become large, soft, sunken areas covered with white fungal growth. Infected fruit often collapse or "melt down" in the field or in storage.

Causal Organisms

Phytophthora blight is caused principally by the soil-borne fungus, Phytophthora capsici, although a similar fungus, Phytophthora parasitica, has also been reported to cause fruit rot on peppers and cucurbits. Both of these fungi are dependent on free water in soil for infection.

Because of this, initial infection by Phytophthora usually occurs on plants growing in poorly drained areas of fields. The fungi that cause Phytophthora blight survive as thick-walled, resistant spores (oospores) in the soil and as mycelium in infected plant tissues. Once introduced, these fungi can survive up to 15 months in moist soil in the absence of host plants. They can also be carried on seed or transplants.

Once stem infection occurs, the fungi produce spores on infected stem tissues, which are then carried by splashed rain onto nearby plants. Lower branches of adjacent plants can also be infected by rain-splashed soil contaminated by run-off water. Fruits in contact with the soil are especially prone to infection.

Spores are produced on newly infected fruit and stems, and new infections can develop quickly in a short time. Warm (68 to 75 degrees F), wet weather is most favourable for disease development and spread of the pathogen. Disease progress declines when dry weather returns.

- Use only certified disease-free seed or transplants.
- Produce pepper plants on raised beds to retard initial stem infection. Arrange beds to prevent puddle formation between rows. Avoid depressions on the surface of the beds where water can accumulate, especially around the base of plants. Avoid planting peppers or cucurbits in poorly drained fields.
- Practice crop rotation so that peppers or cucurbits are grown only every 3 to 4 years to reduce the amount of Phytophthora present in the soil. Use non-susceptible crops such as corn, small grains, beans, crucifers, or potatoes in the rotation.
- Application of fungicides may reduce disease development and spread if used in combination with cultural practices such as crop rotation and raised beds. Application at planting and two or more (depending on the fungicide used) applications after planting are required.

CLUBROOT OF CRUCIFERS

Clubroot is a world-wide problem in temperate climates in the production of cruciferous vegetables such as cabbage, broccoli, cauliflower, radishes, and turnips; and field crops such as mustard and rape. The disease was known as early as the 13th century in England where it was called "finger and toe" disease because of the shape of infected roots.

Symptoms

The most striking symptom of clubroot is an abnormal enlargement of the root system with clubs often thickest at the centre, tapering spindle-like towards the ends. In radishes, clubroot causes distorted swellings on the base of the bulb and along the tap root. In severe cases, entire plantings are destroyed.

Clubroot-infected plants often wilt on sunny days and permanent wilting may accompany advanced decay of infected roots. Severe stunting may be evident if infection occurs early and the disease progresses rapidly. The malformed and greatly enlarged roots are the key symptom of this disease.

Causal Organism

Clubroot is caused by the soil-borne fungus, Plasmodiophora brassicae, which only infects plants in the crucifer family. It infects susceptible host plants through root hairs. Once in the tissue, it stimulates abnormal growth of affected parts, resulting in a swollen club. Infection is favoured by excess soil moisture and low pH, although it can occur over a wide range of conditions.

Once a plant is infected, numerous resistant spores of the fungus are produced in the "clubbed" tissues. As these tissues decay, spores are released into the soil where they can remain infectious for at least 10 years.

Contaminated soil moved by wind or water can serve as a source of infestation of nearby fields causing outbreaks of disease in areas where susceptible crops are planted for the first time. Numerous races of the pathogen have been identified.

Clubroot is an very difficult disease to manage, and heavily infested areas may have to be abandoned for crucifer production. Some control may be achieved with the following measures:

- A good crop rotation programme, growing crucifers on the same soil no more than every third or fourth year, is essential to retard development of a large population of spores on land not already heavily infested.
- Since clubroot is favoured by a low pH, liming soil to pH 7.2 or above may be helpful. Raising soil pH too high, however, may interfere with the growth of succeeding crops other than crucifers. Calcitic lime is usually preferable to dolomitic lime, except for soils low in magnesium, where dolomitic lime is more effective. In course-textured soils, increasing the pH can result in boron deficiency. This may be alleviated by application of boron in transplant water or as a foliar spray.
- With transplanted crops, the use of pathogen-free seedbeds and uninfected plants is essential to prevent introduction of the disease.
- Application of a fungicide in transplant water or rototilled in a band prior to planting may help to reduce disease development.
- Clean and disinfect all machinery before moving it from infested to non-infested land.
- Some resistant cultivars are available. However, plant resistance has not been very useful in clubroot control because of rapid development of new races of the fungus.

COMMON SMUT OF CORN

Corn smut is an extremely common disease of sweet, pop, and dent corn and throughout the world. It is usually not economically important, although in some years yield losses in sweet corn may be as high as 20 per cent.

In Mexico, immature smut galls are consumed as an edible delicacy known as cuitlacoche, and sweet corn smut galls have become a high value crop for some growers in the NE United States who sell them to Mexican restaurants.

Symptoms

The corn plant may be infected at any time in the early stages of growth, but becomes less susceptible after formation of the ear. Above-ground parts may be infected, but it is more common to see smut galls on the ears, tassels, and nodes than on the leaves, internodes, and aerial roots.

The smut gall is composed of a great mass of black, greasy, or powdery spores enclosed by a smooth white covering of corn tissue. The gall may be 4-5 inches in diameter. When leaves are infected, small pustules develop, usually on the midrib, causing some leaf distortion. After the spores mature, the outer covering becomes dry and brittle, breaks open, and the spores sift out. Greatest yield losses occur when the ear becomes infected or if smut galls form on the stalks immediately above the ears.

Causal Organism

Corn smut is caused by the fungus, Ustilago zeae, that survives as a resistant spore in the soil over winter, and possibly for 2 to 3 years. These spores, called teliospores, can be blown long distances with soil particles or carried into a new area on unshelled seed corn and in manure from animals that are fed infected corn stalks.

The teliospores germinate in moist air and give rise to tiny spores called sporidia. The sporidia bud like yeast, forming new spores that germinate in rain water that collects in the leaf sheaths. This leads to infections that are visible in 10 days or more. Wounds from various injuries provide points for the fungus to enter the plant.

The smut fungus is sensitive to temperature and moisture changes. In a warm season, the amount of smut is related closely to the amount of soil moisture, especially during June. When temperatures are lower than normal, there may be little smut even though soil moisture remains high.

- Seed treatment is of no value because few spores are on the corn seed.
- For the home garden, remove smut galls before they break open and bury or burn them. This must be done on a community basis, however, in order to be effective. Smut gall removal is not practical in commercial production.
- In order to reduce infection points from insect injury, control corn borers as first tassels appear by application of insecticides when insect populations are high.
- Avoid injury of roots, stalks, and leaves during cultivation.
- Deep plow diseased corn stalks in the fall to bury surviving spores.
- Use resistant hybrids or varieties. Dent corn is generally more resistant than sweet or popcorn. In sweet corn, the larger, later-growing varieties usually are more resistant than the smaller, early varieties.

BACTERIAL WILT OF CUCURBITS

After vine crops begin to run, gardeners and farmers often notice individual leaves with severe wilt symptoms on sunny days. Within a week or two the condition spreads to entire vines which do not recover from the wilt. This disease, called bacterial wilt, is especially common with cantaloupes and cucumbers. Squash and pumpkins may not wilt as rapidly, but may be dwarfed with extensive blossoming and branching. Watermelons are rarely affected.

Symptoms

Wilting of individual leaves or vines of the plant is the characteristic symptom. One or a few leaves wilt and become dull green. The disease spreads from the leaves downward into the petioles and then the stem until the entire plant wilts and dies.

There are other factors, such as vine borers and soil-borne fungal pathogens, that may cause cucurbits to wilt. Sometimes, if an affected stem is cut off near the ground, the sap may be milky in appearance or sticky and, if touched with the finger, the sap will string up to half an inch. This is a helpful test in diagnosis of bacterial wilt, but cannot be depended upon for positive identification.

Causal Organism

This disease is caused by a bacterium, Erwinia tracheiphila, that overwinters in the bodies of the striped and 12-spotted cucumber beetles. In the spring, the beetles emerge from the ground and feed on young plants, introducing bacteria into the leaves or stems. The bacteria reproduce in the water-conducting vessels, producing gums that interfere with water transport.

The beetles and bacteria are so intimately related that controlling the beetles will control infection by the bacteria. Once infection has occurred, however, no control is possible and wilting plants should be removed, if practical.

Management

The only practical management measure is to use an insecticide when seedlings first emerge to control the black and yellow cucumber beetles. Early infections are most severe, but total control depends on applications continuing at frequent intervals during the growing season.

In some cases, if insect pressure is heavy, it may be necessary to apply an insecticide when plants are just cracking the soil, but have not yet emerged. Management of this disease is completely linked with preventing feeding of cucumber beetles on susceptible hosts.

FUSARIUM AND VERTICILLIUM WILTS

Solanaceous crop plants (tomato, potato, pepper, and egg- plant) may be infected at any age by the fungi that cause Fusarium wilt and Verticillium wilt. The wilt organisms usually enter the plant through young roots and then grow into and up the water conducting vessels of the roots and stem. As the vessels are plugged and collapse, the water supply to the leaves is blocked.

With a limited water supply, leaves begin to wilt on sunny days and recover at night. Wilting may first appear in the top of the plant or in the lower leaves. The process may continue until the entire plant is wilted, stunted, or dead.

Tomato and potato plants may recover somewhat but are usually weak, unthrifty, and produce fruit of low quality. Peppers typically collapse rapidly and die. Fusarium and Verticillium wilts are rarely significant in field grown tomatoes due to the widespread incorporation in tomato cultivars of genes for resistance to the pathogen.

However, the resurgent interest in planting "heirloom" tomato varieties which do not carry resistance genes has resulted in increased incidence of Fusarium and Verticillium wilts. Additionally, new races of both pathogens have been identified that are capable of overcoming the resistance in many popular tomato varieties.

Verticillium race 2 is now common in tomatoes, but its importance in reducing yield is not known at this time. There is very little genetic resistance available to either disease in pepper or eggplant.

Symptoms

Fusarium Wilt

Fusarium wilt symptoms begin in tomato and potato as slight vein clearing on outer leaflets and drooping of leaf petioles. Later the lower leaves wilt, turn yellow and die and the entire plant may be killed, often before the plant reaches maturity. In many cases a single shoot wilts before the rest of the plant shows symptoms or one side of the plant is affected first. If the main stem is cut, dark, chocolate-brown streaks may be seen running lengthwise through the stem.

This discolouration often extends upward for some distance and is especially evident at the point where the petiole joins the stem. Potato tubers may show browning of the vascular ring as well as browning at the stem end and decay where stolons are attached. In pepper, lower leaves do not begin to wilt until roots and the base of the stem have already started to decay. Wilting of the entire plant soon follows. Dark brown, sunken, and eventually girdling cankers may be seen at the base of the pepper plant. In eggplant, wilting progresses from lower to upper leaves, followed by collapse of the plant.

Verticillium Wilt

Verticillium wilt symptoms on tomato, potato, and eggplant are similar to those of Fusarium wilt. Often no symptoms are seen until the plant is bearing heavily or a dry period occurs. The bottom leaves become pale, then tips and edges die and leaves finally die and drop off. V-shaped lesions at leaf tips are typical of Verticillium wilt of tomato. Infected plants usually survive the season but are somewhat stunted and both yields and fruits may be small depending on severity of attack. A light tan discolouration in the stem similar to that caused by Fusarium wilt may be found but is usually confined to lower plant parts. The discolouration is typically lighter in colour than with Fusarium wilt. Symptoms on one side of the plant only are sometimes seen.

In potatoes the pathogen may be part of a complex that includes, among others, the root lesion nematode and the bacterial soft rot organism, resulting in premature plant death ("potato early dying disease"). Tubers from Verticillium-infected plants may show light brown vascular discolouration, usually restricted to the stem end. In pepper, the lower leaves wilt, then leaf tips and margins dry and turn brown. Brown streaks in the vascular tissue can be observed well up into the plant, which rapidly collapses and dies.

Causal Organisms

Fusarium wilt in solanaceous crops is caused by several different types of the fungus Fusarium oxysporum. These are: F. oxysporum f. sp. lycopersici (tomato), F. oxysporum f. sp. melongenae (eggplant) and F. oxysporum var. vasinfectum (pepper). Fusarium wilt in potato is caused by a complex of up to four different Fusarium spp. All of the Fusarium wilt pathogens are generally specific to their hosts and are soilborne. They are warm weather organisms, and therefore Fusarium wilts are most serious later in the growing season.

Verticillium wilt is caused by the fungi Verticillium albo-atrum and V. dahliae. These fungi attack a wide range of plant species, including cultivated crops and weeds. They are soilborne in field and greenhouse soils where they can persist for many years. V. albo-atrum is a cool weather organism that grows best when soil temperatures are between 65 and 75 degrees F. V. dahliae is more active between 75 and 83 degrees F.

Although disease is retarded by the higher temperatures that favour Fusarium wilt, visible symptoms may appear to be more severe when high temperatures exist, due to restricted water movement in the plant brought about by damage done to the water conducting vessels earlier in the growing season.

- Because Fusarium and Verticillium fungi are widespread and persist several years in soil, a long crop rotation (4 to 6 years) is necessary to reduce populations of these fungi. Avoid using any solanaceous crop (potato, tomato, pepper, eggplant) in the rotation, and if Verticillium wilt is a problem, also avoid the use of strawberries and raspberries, which are highly susceptible. Rotate with cereals and grasses wherever possible.
- Keep rotational crops weed-free (there are many weeds hosts of Verticillium).

- Whenever practical, remove and destroy infested plant material after harvest.
- Maintain a high level of plant vigour with appropriate fertilization and irrigation, but do not over-irrigate, especially early in the season.
- Plant disease resistant tomato varieties, labeled V (for Verticillium) and F (for Fusarium). These disease resistance designations are usually shown in seed catalogues. Fusarium- or Verticillium-resistant varieties of eggplant, potato, and pepper are generally not available.
- If soils are severely infested, production of solanaceous crops may not be possible unless soil fumigation is an option.

BACTERIAL SPOT OF PEPPER

Bacterial spot of pepper is one of the most destructive diseases of pepper in climates where high temperature and frequent rainfall occur during the growing season. The disease causes spots on leaves and fruit; leaf defoliation; and a reduction in plant growth, fruit yield, and quality. Bacterial spot is also a serious problem on tomato, although not all strains of the pathogen can cause disease on both hosts.

Symptoms

Characteristic bacterial spot symptoms can appear on the leaves, fruits, stem, and petioles. On leaves, symptoms begin as small, yellow-green circular lesions surrounded by a yellowish halo. These spots appear water-soaked under wet conditions. As the lesions mature, a general yellowing extending from the area around the lesions develops on diseased leaves and the centre of the spots become brown to black and sunken.

Tissue in the centre of the lesion often dries and breaks away, giving a "shot-hole" appearance to the leaf. When spots are numerous, they may join together and form irregular discoloured streaks along the veins and leaf margins. Edges and tips of leaves may die, then dry and break away, causing leaves to appear ragged. Severely spotted leaves turn yellow or brown and fall from the plant; young leaves can be distorted. Fruit spots begin as green, circular, slightly raised lesions which eventually become brown or dark, raised, and about 1/8 inch in diameter. Centres of the spots become necrotic, corky, and scab-like.

On stems and petioles, lesions are elongated and blackened, and can kill leaflets. Cotyledons are particularly susceptible to bacterial spot; lesions are initially small, sunken, and silvery. They later become darker in colour.

Causal Organism

Bacterial spot of pepper is caused by a bacterium, Xanthomonas campestris pv. vesicatoria, which can overwinter in crop residue in or on soil, on or in seeds, and on wild host plants. Pepper seeds infested with the pathogen are a major source of inoculum for bacterial spot as well as the major means of long distance spread of the pathogen. The pathogen can survive on dried seeds for up to 10 years. It is not able to survive free in soil for a long time, but can survive up to 6 months in infected crop debris in soil. The bacterium penetrates leaves through stomata and/or wounds, and fruits through wounds created by wind-driven sand, insect punctures, or mechanical injury. Dissemination of the bacterium occurs between and within fields by water-splashing, aerosols, or during cultivating, hoeing, thinning of direct seeded plants, transplanting, or harvesting.

In the United States, numerous physiological races of the bacterium are known, most of which have been found:

- Use pathogen-free seeds and transplants.
  Use sodium hypochlorite-treated seed to reduce bacterial populations.
- Practice crop rotation with non-host plants such as corn and soybean so that peppers are grown only every 3 to 4 years. However, do not use soybeans in the rotation if white mold (Sclerotinia sclerotiorum) has been a problem.
- Deep plow to bury infected crop debris.
- Avoid working in the field when foliage is wet.
- Eliminate wild host plants such as nightshade and ground cherry in and around field.
- Application of copper-containing pesticides may be helpful for preventing development and spread of bacterial spot.
- None of the currently available pepper varieties are resistant to all known races of the bacterial spot pathogen. However, use of varieties resistant to one or more races of the pathogen may provide some control, depending on the races present.

DOWNY MILDEW OF CRUCIFERS

Downy mildew affects all cultivated plants and weeds in the crucifer family. It can be a serious problem in commercial production of cabbage, broccoli, cauliflower, radish, turnip, mustard, collard, and cruciferous greens. Under favourable conditions, it may cause serious losses in the field or may develop after harvest and cause deterioration of product quality during packing and shipping.

Symptoms

Plants can be infected at any stage of development. In seed beds, cotyledons and primary leaves are invaded resulting in fungal growth visible on the underside of the leaf. Later a slight yellowing develops opposite the fungal growth on the upper side of the leaf.

The young leaf or cotyledon, when yellow, may drop off. Older leaves usually persist and infected areas gradually enlarge, turn bright yellow, then become tan and papery. Rarely the affected leaf may develop hundreds of minute darkened specks. Under cool, moist conditions, a white mildew growth can be seen on the underside of infected leaf lesions. Symptoms may appear on other plant parts as well.

The fleshy roots of turnips and radishes may develop an internal, irregularly shaped discolouration extending from the crown downward. The flesh may be brown or black or show a form of net necrosis. In advanced stages, the skin becomes roughened by minute cracks and the root may split open. In radish these symptoms may be confused with those caused by Rhizoctonia.

Causal Organism

The fungus causing downy mildew in crucifers, Peronospora parasitica, overwinters in roots or in decaying portions of diseased plants. Thick-walled resting spores may form in stems, cotyledons, and other fleshy parts of infected host plants. On growing plants, the fungus produces large numbers of spores that are blown about by wind and splashed by rain.

Moisture and temperature are important in the spread and reproduction of this fungus. High relative humidity during cool or warm, but not hot, periods promotes its growth and sporulation. Presence of a water film on the foliage from fog, drizzling rain, or dew allows spores to germinate, infect, and produce more spores on a susceptible host in as few as 4 days.

- Use a crop rotation plan that excludes production of any type of cruciferous crop for at least 2 out of every 3 years.
- Practice sanitary measures such as the use of clean seed beds away from other crucifer production and the destruction of cruciferous weeds.
- Use a planting site and plant spacing pattern that expose plants to full sun throughout the day.
- If severe disease pressure is expected, apply a registered fungicide weekly beginning soon after emergence.
- Disease resistant cultivars are not available for most cruciferous crops. However, some hybrid cultivars of broccoli are resistant or tolerant to downy mildew.

BLACK ROT OF CRUCIFERS

Black rot is the most serious disease of crucifer crops world wide when environmental conditions (relatively high temperature and humidity) are favourable. The disease affects primarily aboveground parts of plants at any stage of growth and causes high yield and quality losses, especially in tropical and subtropical regions during the rainy season.

All vegetables in the crucifer family, including broccoli, Brussels sprouts, cabbage, cauliflower, Chinese cabbage, kale, mustard, radish, rutabaga, and turnip, are susceptible to black rot. Many cruciferous weeds such as Shepherd's

Purse, wild mustard, and yellow rocket may also be hosts of this pathogen. The characteristic black rot symptom on most cultivated crucifer plants is the appearance of yellow, V-shaped lesions along the margins of leaves. The point of the V-shaped lesion is directed toward a vein.

When lesions enlarge, wilted tissue expands toward the base of leaves. Eventually the diseased areas become necrotic and the veins turn black or brown. The infection may move down the vascular tissue of petioles and then spread up and down the stems. When stems and petioles of an infected plant are cut crosswise or lengthwise, the black-brown vascular tissue with yellowish bacterial slime is observed. These symptoms may be confused with Fusarium yellows, except that Fusarium causes brown vein discolouration without bacterial slime.

Moreover, symptoms of black rot may vary according to age of host, host genus, species, and cultivar and even environmental conditions. Symptoms on cauliflower may appear as numerous black or brown specks, scratched leaf margins, black veins, and discoloured curds. Many cruciferous weed species do not exhibit any of these characteristic symptoms even when infected.

Symptoms

Causal Organism

Black rot of crucifer is caused by a bacterium, Xanthomonas campestris pv. campestris (Xcc). The bacteria can overwinter in plant debris, in and on seeds from diseased plants, and in and on weeds. The pathogen may survive in diseased crop residue buried in soil for up to 2 years, but not more than 60 days free in soil.

The major source of these bacteria is infected seeds, which enable long-distance spread of the disease. The pathogen is spread within and between fields by splashing water, wind, insects, machinery, and irrigation or drainage waters. The bacteria infect the cotyledons and young leaves through natural plant openings (stomata, hydathodes) or wounds and then migrate between cells until they reach the xylem tissue where they spread throughout the plant. Free moisture is required for infection by the pathogen. After infection, symptoms may appear on plants within 7 to 14 days under optimum conditions (25 to 30 degrees C).

Effective management of black rot of crucifers depends on the application of the following practices in combination:

- Use black rot-tested, disease-free seed grown in an arid production area.
  - If source of the seeds is unknown, or infested seedlots must be used, treat seed with hot water to eradicate pathogenic bacteria. Cabbage, broccoli, and Brussels sprouts can be treated at 50

degrees C for 25 minutes, while seeds of cauliflower, kale, turnip, and rutabaga are treated for 15 minutes. However, this treatment may reduce the viability of seed. Therefore, some other chemical seed treatments, including, sodium hypochlorite, hydrogen peroxide, and hot acidified cupric acetate or zinc sulfate can be applied to eliminate the bacteria from crucifer plant seeds.

- Use certified disease-free transplants.
- Practice crop rotation where crucifers are grown only every 3 to 4 years to eliminate the inoculum sources from diseased crop debris in the soil.
- Good sanitation practices should be performed to prevent disease spread.
  - Eliminate all volunteer crucifer plants from previous crops and alternative wild host plants within and around the field.
  - Do not apply manure that may contain crucifer residues.
  - Do not use sprinkler irrigation.
  - Avoid working in the field when plants are wet.
  - Do not allow machinery and equipment movement from infested areas to non-infested fields.
  - Deep plow to bury all crucifer residues after harvest.
- Application of fixed copper pesticides in the field may help to reduce spread of the disease.
- A few black rot-resistant cultivars of cabbage and other crucifers are commercially available. These resistant cultivars should be used in crucifer growing regions where black rot is a common problem.

GUMMY STEM BLIGHT AND BLACK ROT OF CUCURBITS

Gummy stem blight is an important disease of squash, pumpkins, cucumbers, watermelons and other field-grown cucurbit crops. It can occur at any growth stage, from seedlings to mature plants. This disease on fruits, in the field or in storage, is called black rot. The disease also can cause extensive damage to all above-ground parts of greenhouse-grown cucumbers.

Symptoms

Gummy stem blight occurs on all plant parts except roots. Leaf symptoms appear as dark yellow or reddish-brown lesions in various shapes. Lesions begin at leaf margins and extend rapidly back into the leaf blade, causing curling, shriveling, and death of the entire leaf. Pimple-like structures (pycnidia) may be found in leaf lesions by close inspection with a hand lens.

Fruit symptoms vary among crops. Winter squash (Hubbard, butternut, etc.) are likely to show symptoms primarily on fruit or older leaves. Black rot symptoms appear as a brown to black rot of the rind, flesh, and seed cavity accompanied by heavy white and black fungus growth. Lesions may develop anywhere on the fruit, first as water-soaked areas dotted with pycnidia that ooze yellowish masses of spores.

On Hubbard-type squash, the brownish-black rot extends down into the flesh and seed cavity. Seeds often become dotted with small black pycnidia. On butternut squash, lesions are brown with irregular ring patterns and are superficial over the skin surface, not penetrating into the flesh or seed cavity.

Infection usually occurs in the field, causing water-soaked, cracked, brownish cankers on the vines. A reddish gum may develop in these cracks, although this alone is not a diagnostic sign. (Fusarium and scab may also produce a reddish gum.) Fruit may decay at the site of attachment as a result of the fungus invading the stem.

Butternut squash and gourds may develop black rot before harvest, but Hubbard squash are resistant during growth and do not show symptoms until the storage period. Fruit rot on greenhouse cucumbers usually begins at the blossom end of immature fruit. Lesions are firm and become dark brown to black when cut open. Occasionally, lesions develop on one side of a fruit, causing it to hook as it grows.

Causal Organism

Gummy stem blight is caused by the fungus Didymella bryoniae. The pathogen can be seed-borne or can survive on organic debris from previously infected cucurbits or on wild or volunteer cucurbits. The gummy stem blight fungus produces two types of spores. Windblown ascospores are likely to start the disease in a field.

Later, pycnidiospores are released in a gummy substance that makes them adaptable for short distance spread by splashing water. Spore production and infection are influenced by moisture and temperature. A moisture film from dew, rain, or overhead irrigation is necessary for spore germination. Optimum temperature for infection is 61 to 75 degrees F. Low night temperatures, particularly in greenhouses, may cause water droplets to exude from leaf points and condensation to form on leaves, favouring infection by the fungus at those points.

Infection of fruits commonly occurs through harvest wounds or through dying flowers:

- Use only disease-free seed produced in arid western locations.
- Plow crop refuse deeply to reduce survival of the fungus as soon as crop is harvested.
- Practice crop rotation with non-cucurbit crops so that cucurbits are grown only every 3 to 4 years.
- Apply foliar-protectant fungicides on a routine basis.
- Avoid wounding fruit during harvest and store fruit at 45 to 50 degrees F to prevent postharvest fruit rots.
- Resistant cultivars are not currently available.

MUMMY BERRY OF BLUEBERRY

Mummy berry is one of the most serious diseases of blueberry. Once the disease becomes established in a planting, it can destroy most of the crop. Losses result from:

- Rotted berries; and
- Killing or blighting of blossoms, blossom and leaf clusters and young shoots.

Symptoms

Blighting of new shoot tips and blossoms can be easily mistaken for frost damage. By blossom time, the infected young leaf and shoot growth will wilt, turn brown and die. About a week or so after infection of the early new growth occurs, dead areas develop on the petioles and along the midrib and veins of the leaves or at the base of flowers. Berries that develop from infected flowers may attain nearly full size before turning tan or gray and shriveling into hard mummies, which drop to the ground at or before harvest.

Causal Organism

Mummy berry is caused by the fungus, *Monilinia vaccinii-corymbosi*. The fungus overwinters in the shriveled mummies on the ground. In early spring, cup- or globe-shaped structures of the fungus called apothecia are produced on mummified berries during cool rainy periods. Spores produced inside apothecia are released into the air and carried by the wind to young developing leaf shoots and flowers where they cause primary infections. If moisture is not present, the fungus will not produce spores. However, the fungus may survive in mummies for one year or more. Another type of fungus spores (conidia) is produced on dead tissue that results from primary infections. These conidia are spread during bloom by wind and insects, and result in secondary infection of flowers. Fruit that develops from infected flowers turns into mummies and falls to the ground.

Control

Removing mummified berries from the planting will greatly aid in controlling the disease. Removing these berries is not practical on a commercial scale, but may be of value in backyard plantings where there are just a few plants. After removing mummies, burn or bury them. Cultivation in early spring to disturb or cover the mummies has been reported to be effective. Mummies that are disturbed or covered with soil at this time remain dormant or do not produce spores. Cultivation between rows and raking under plants to disturb or cover mummies should be done as early as possible in the spring and repeated after each hard rain until after bloom. If just a few mummies are missed, they can produce enough spores to infect the planting. Where mummy berry is a problem, a good fungicide spray programme is essential.

LEATHER ROT OF STRAWBERRY

Leather rot of strawberry has been reported in many regions of the United States. In many areas, it is considered a minor disease of little economic importance. However, excessive rainfall during May, June, and July can lead to severe losses in fruit yield and quality resulting from leather rot. Commercial growers have lost up to 50 per cent of their crop to leather rot. The leather rot fungus primarily attacks the fruit, but many also infect blossoms.

Symptoms

Leather rot can infect berries at any stage of development. Where the disease has been a problem, infection of green fruit is common. On green berries, diseased areas may be dark brown or natural green outlined by a brown margin. As the rot spreads, the entire berry becomes brown, maintains a rough texture, and appears leathery. The disease is more difficult to detect on ripe fruit.

On fully mature berries, infection may result in little colour change or discolouration ranging from brown to dark purple. Infected ripe fruit are usually softer to the touch than healthy fruit. When diseased berries are cut crosswise, a marked darkening of the water-conducting system to each seed can be observed. In later stages of decay, mature fruits also become tough and leathery. Occasionally, a white moldy growth can be observed on the surface of infected fruit. In time, infected fruit dry up to form stiff, shriveled mummies.

Berries that are affected by leather rot have a distinctive, unpleasant odor and taste. Even the healthy tissue on a slightly rotted berry is bitter. This presents a special problem to growers in pick-your-own operations. An infected mature berry with little colour change may appear normal and be picked and processed with healthy berries. Consumers have complained of bitter tasting jam and jelly made with berries from fields where leather rot was a problem. Leather rot is observed most commonly in poorly drained areas where there is or has been free standing water, or on berries in direct contact with the soil.

Causal Organism

Leather rot is caused by the fungus, *Phytophthora cactorum*. The fungus survives the winter as thick-walled, resting spores, called oospores, that form within infected fruit as they mummify. These oospores can remain viable in soil for long periods of time. In the spring, oospores germinate in the presence of free water to produce a structure called a sporangium. A second type of spore, called a zoospore, is produced inside the sporangium. Up to 50 zoospores may be produced inside one sporangium. The zoospores have tiny flagella and can swim in a film of water. In the presence of free water on the fruit surface, the zoospores germinate forming a germ tube and infect the fruit. In later stages of infection, sporangia are produced on the surface of infected fruit under moist conditions.

SEED-BORNE VIRAL DISEASES

In general, viruses are not transported through true seed or pollen, but at least 30 viruses including the potato spindle tuber viroid are trans witted in this way. In most virus/host combinations, the proportion of seedling infected through the seed is very less. On the other hand, in tobacco ring spot virus in soybean, most of the seed may be infected (Desjardins *et al.*, 1954). Seed transmission provides a very effective means of introducing a virus into a crop at an early stage, giving randomized foci of infection throughout the planting. Viruses may stay in seeds for long periods so that the commercial distribution of a seed-borne virus over long distances may be very effective. Fulton lists 36 viruses known to be transmitted internally through the seed. These include tobacco ring spot, lettuce mosaic, barley stripe mosaic, bean mosaic, southern bean mosaic, elm mosaic and *Prunus* ring spot.

Some interesting studies have been made on the seed transmission of the virus responsible for false stripe disease of barley. Particles *of* the virus have been found to be associated with the disease. These occur in large numbers in the leaves, embryos, endosperms, pollen and unfertilized pistils. The presence of the rods in seed produced from healthy pistils pollinated by pollen from diseased plants suggested that the disease is seed transmitted. Seeds from the diseased pollen of diseased plants and healthy pistils produce a small percentage of diseased seedlings (Gold *et al.*, 1954). There are examples of a virus being transported through the seed of one host species but not through the seed *of* another. Cucumber mosaic virus is transmitted through the seed of the wild cucumber *(Micrampelis lobata)* but pot through the seed of the cultivated cucumber. Cow pea mosaic virus is transported through the seeds of asparagus bean (a variety of *Vigna unguiculata)* but not through the seeds *of* cow pea *(Vigna sinensis).* Tobacco ring spot virus is carried through the seeds of *Petunia* sp. (Henderson, 1931). There are some other examples *of* transmission of viruses through the seeds *of* other host plants. Certain viral diseases are seed transmitted with the help of nematode borne viruses (Murant and Lister, 1967).

It is not known why fewer virus diseases are seed transmitted. Viruses which are limited to the phloem region cannot enter the seed because the phloem of the mother plant and the embryo are not connected directly. With other viruses the possible explanation is that they cannot replicate and/or are degraded in meristematic tissues as is found in developing seeds. Weathers and Calavan (1959) showed that the seedlings grown from both true seed and nucellar seed of infected citrus are virus free. In the case of southern bean mosaic, the virus is present in the embryos of immature seeds of *Phaseolus vulgaris.* The virus can come from either the pollen *or* ovule-parent but is inactivated as soon as the seed matures and is not transmitted. Seed-borne viruses can be detected by inoculating extracts of embryos or pollen into indicator plants. The particles of barley stripe mosaic virus can be found in extracts of individual seeds of barley by electron microscopy *or* microserological tests (Hamilton; 1965).

A large number of viruses are transported through vegetative parts. Thus, any vegetative part taken for propagation, such as tubers, bulbs, corms, runners and cuttings spreads the virus if the host plants from which these have been taken were infected.

SEED-BORNE FUNGAL DISEASES

A large number of fungal pathogens are spread through the seed and the vegetative propagating parts. Some of the fungal pathogens transmitted in this way are many cereal smuts, leaf blight of wheat caused by *Alternaria triticina,* leaf stripe and seedling blight of oats *(Helminthosporium avenae),* leaf stripe of barley *(H. gramineum),* anthracnose of French and summer beans *(Colletotrichum lindemuthianum),* red rot of sugarcane *(Colletotrichum falcatum),* seedling blight of flax *(Colletotrichum linicolo),* and late blight of potato *(Phytophthora infestans).* These are some important examples where the pathogens are transmitted on the seed surface, inside the seed coat or deeper in the embryo. A list of the important fungal, bacterial and viral pathogens prevalent in India which are transmitted through the seed is given in Table. Seed infection by fungi," bacteria and viruses may arise in various ways, such as by floral infection, systemic invasion through the vascular system of the 'plant' to the seed, direct penetration through pods or other fruits and invasion of the seed during storage and harvesting.

In the case of *Ustilago nuda* and *U. tritici,* recent researches have been done on floral infection in the field and the pathway of infection of the seed. Invasion usually takes place through the ovary wall and not via the stigma and style as was previously supposed. The hyphae cross the cells of the pericarp (fruit coat), enter the testa (seed coat) and then move towards the ovary. After reaching the ovary the hyphae turn sharply from the testa, enter the scutellum and pass into the growing point of the embryo. Mycelium is intracellular within the pericarp and testa but it becomes intercellular within the tissues of the embryo. There is apparently no effect on grain development and it is not possible to know from external appearance that the seed is infected. An excellent technique has resulted from the researches of Simmonds (1946) and Popp (1958). In Popp technique embryos are separated from the seed by boiling the seeds in a mixture of 3 per cent (w/v) sodium hydroxide and 12 per cent (w/v) ethanol containing 15 per cent (w/v) detergent. The embryos are then boiled in a 12 per cent (w/v) ethanol containing 15 per cent (w/v) sodium hydroxide and heated in a 3:1 mixture of ethanol and glacial acetic acid. This is done to clear the tissues and to remove oil from the embryo. The embryos are then heated in 45 per cent (v/v) acetic acid containing 0.1 per cent trypan blue to stain the fungus hyphae. The ,embryos are then heated and mounted in 45 per cent (v/v) lactic acid.

SEED TREATMENT

Chemical Methods for Plant Disease Control

The annual loss to world crops as a result of disease has been estimated at 25,000 million dollars; of this, a major part is due to fungal pathogens. Surface deposits of fungicides can control many fungal diseases if properly applied, but it appears that more effective control can be achieved with systemic fungicides which can penetrate to remove pests from seeds.

There is an enormous amount of literature available on chemicals used for plant disease control. But the following are worth mentioning: *Principles of fungicidal action* by Horsfall (1956); *Fungicides, An Advanced Treatise* in two volumes by Torgeson (1967, 1969); plant diseases and their chemical control discussed by Evans (1968); chemistry and physiology of fungicidal action by Owens (1963); fungicidal chemistry by Rich (1960); physiology of fungitoxicity by Sisler and Cox (1960); mechanism of toxicity of agricultural fungicides by Somers (1962); soil fungicides by Domsch (1964); fate of fungicides in plants by Sijpesteijn and Kirk (1965); adaptation of fungi to metal toxicants by Ashida '(1965); synergism among fungicides by Anna Scardavi (1966); effect of fungicides on protein and nucleic acid synthesis by Sisler (1969); mechanism of action of antifungal antibiotics by Gottleib and Shaw (1970); role of fungicides in modernizing agriculture in India by Mehta (1971); *Chemical Control of plant Disease: An Exciting Future* by Sbragia (1975); and acquired resistance to fungicides by Dekker (1976). A comprehensive and up-to-date reference source available in the field of fungicides is *Antifungal Compounds,* in two volumes edited by Siegel and Sisler (1977).

CHEMICALS USED FOR SEED TREATMENTS

- *Thiram* (Tetrametltyl thiuram disulphides) TMTD

  TMTD is a condensation product of dithiocarbamic acid (2 molecules in which the H atoms have been substituted by methyl groups). This was used as an activator in the production of synthetic rubber.

  Thiram is being sold under a variety of other trade names, such as 'arasan', 'tundas', 'terson', 'tulisan', etc.

  It is a leading seed treatment for wheat, rice, gram, peas, mustard, linseed, sorghum and onion smut. Thirani is. also used in the control of foliage diseases and acts as an insecticide. Among the soil-borne pathogen controlled by it are *Pythium, Rhizoctonia, Fusarium* and *Protomyces.*

As a foliage spray, thiram was not as successful as its metallic compounds. For this reason, several compounds have been developed.

- *Ferbam* (Ferric dimethyl dithiocarbamate)
  In ferbam, three molecules of dithiocarbamic acid react with one atom of iron. Its trade name is fermate or ferbam, karbam or coromet. Ferbani is compatible with most pesticides, but its use with copper, lime or mercury compounds tends to reduce fungicidal efficient Ferbam has been successfully used as a protectant fungicide, against a wide variety of fungus diseases of fruits and vegetables. As a foliage spray it is also used to correct iron deficiency.
- *Ziram* (Zinc dimethyl dithiocarbamate)
  The trade names are cuman, zerlate, karbam white, corozate, etc. In India, this fungicide is available under the trade name of Cuman. It is used against many foliar diseases in field and orchard crop.
- *Nabam* (Disodium ethylene bisdithiocarbamate)
  This is sold as Dithane 0-14, Parzate liquid or Dithane A-40. Nabam has been found effective against many leaf blights, such as early and late blight of potato and seedling blights caused by *Pythium. Rhizoctollia* and *Fusarium* species.
  A similar compound of zinc was later developed. This is zineb.
- *Zineb* (Zinc ethylene bisdithiocarbamate)
  Zineb is derived by the substitution of sodium in nabam with zinc. It is called Ditballe Z-78, Parzate C or Lonacol. Zineb is mainly used for foliar sprays against diseases, such as early and late blight of potatoes and' tomatoes, blast of rice, leaf blight of rice, ripe rot of chillies and downy mildew of maize.
- *Maneb* (Mangnese ethylene bisdithiocarbamate)
  This is called as Dithane M-22, Manzate or Oithane M-4S. Maneb is excellent for the control of foliage and fruit diseases *of* many vegetables, flowers, trees; turf and some fruits in many diseases maneb has been found more effective than zineb. It is especially used in vegetable crops as a spray material. Some of the diseases controlled by zineb are bean anthracnose caused by *Colletotrichum lindemuthianum,* downy mildew and anthracnose of cucurbits, fruit rot of chillies, citrus greasy spot caused by *Cercospora* spp., lettuce downy mildew, maize leaf blight, blight of tea caused by *Exobasldium vexans,* and leaf spots and blights caused by *Alternaria, Phytophthora* and *Cercospora* on crops, such as potato, tomato, cabbage, carrot, beet and cauliflower.
  - *Chloranil* (Tetrachloro-p-benzoquinoe)
    It was reported in 1940 by Cunningham and Sharvelle to give protection. Chloranil is mainly used in seed treatment. It is sold

as spergon and is used in seed and bulb treatment of, legume flowers and vegetables for seed treatment.

- *Dichlone* (2,3 dichloro-l,4 naphthaquinone)

  It is one of the later invention, this chemical is used mainly for seed treatment.

  It has been reported that this was four 'to eight times more effective as chloranil in the protection of legume and cotton seed.

- *Phenols*

  Phenols have shown exceptional qualities as bactericides, but as fungicides their use has been limited. Phenols have been chiefly applied to textiles and woods as preservative fungicides', but have also been used as soil fungicides and in seed treatments. Derivatives of cresol and O-phenyl phenol have been mainly used as preservatives. The bisphenols are a particularly interesting group of phenolic fungicides. Horsfall and Rich (1950) reported the fungicidal potential of thiobisphenols.

3

Seed Testing and Sampling

SEED TESTING

The International Seed Testing Association (ISTA) and the International Organization for Standardization (ISO) have developed standards for sampling lots of seed and grain, respectively. ISO terminology refers to primary samples as increments, the composite sample as the bulk sample and the submitted sample as the laboratory sample. In this report, sampling terminology and sampling protocols are based on ISTA standards for discussion on seed-for-sowing and on ISO standards for discussion on grain.

Based on ISTA terminology, the sampling process involves taking a number of primary samples from the original seed lot. The number of primary samples required depends on the size of the seed lot and how well it is mixed. If adventitious GM material is unevenly distributed throughout the lot then more primary samples are required to accurately represent what is in the lot. The primary samples are combined and mixed to form a composite sample which is representative of the whole seed lot. As the concentration of GM material in the lot decreases, larger composite samples are required to ensure that the composition of the lot is accurately reflected. If the composite sample is too large to be sent to the laboratory, it is reduced in size (with very thorough mixing) into a submitted sample. In the laboratory, the submitted sample is mixed and further sub-divided into one or more working samples which are then analysed using the chosen test method.

SEED SAMPLING

Australia imports seed and also grain for food and feed from countries that commercially produce GM varieties. Using the example of imported labelled non-GM canola seed for breeding purposes; it could be tested for the presence of both GMOs that are approved in Australia for commercial release, and those GMOs that may be unapproved in Australia but are approved in the countries exporting canola seed to Australia. Testing for unapproved GMOs in imported seed for breeding in Australia helps ensure sown seed is compliant with the Australian regulatory system; dealings with GMOs are illegal in Australia unless authorised under the *Gene Technology Act 2000* (Cwlth). If imported seed meets regulatory requirements, and there is sureness that unapproved GM canola was not being bred and sown, then AP testing further along the supply chain could focus on testing for only Regulator-approved GMOs in marketed non-GM grain to enable market requirements to be met. In the case of GM canola approved for sowing in Australia, labelling thresholds for AP of GM events in non-GM canola grain are lower than or equal to equivalent thresholds in the countries to which Australian canola is exported. Hence, if the non-GM canola meets Australian thresholds, then it may not require re-testing to ensure it meets the receiving country's standards.

For cotton, as the majority of cotton grown in Australia is GM, testing for the AP of GM events in cottonseed is required only in the case of certified non-GM cottonseed that is being exported to niche markets. For soybean and maize, no GM events have been approved by the Regulator for commercial release.

HOW SHOULD SAMPLING BE DONE

The strategies and techniques used in taking, mixing and reducing the samples are designed to ensure that the composition of the working sample analysed in the laboratory corresponds as closely as possible to the composition of the original lot. In reality this process does not work perfectly, and the errors associated with each of the sampling steps need to be considered when estimating the total uncertainty of the sampling process. Two linked factors are involved in developing strategies for sampling and testing. The first is the level of confidence required. This is a measure of how confident one can be that a calculated result is correct. A confidence level of 100 per cent means that 100 out of 100 times, the result given is correct. A confidence level of 95 per cent means that 95 out of 100 times, the result is correct. Also to be taken into account is the relative uncertainty of the measurement (often expressed as the error). Uncertainty is given as a projection of how far away the actual value could be from the calculated result due to the cumulative errors of the processes leading to that result. It is expressed as a range around a value (*i.e.* 1 per cent GM ± 0.05, or 1 per cent GM with 5 per cent relative uncertainty).

The application of confidence levels and uncertainty to AP testing plays a very large role in deciding the sample size required for testing. Statistics can identify the minimum bulk sample size required to provide a specified level of confidence that the GM content of the sample lies within a specified error range of the actual composition of the lot. An 'ideal' homogenous seed lot, larger numbers of seeds are required as the proportion of GM content in the lot decreases. In the case of a perfectly homogenous lot, only a single increment of the requisite size would be needed. However, as it cannot be assumed that real lots are homogenous, most sampling protocols specify the collection of multiple increments from different parts of the lot with the aim of obtaining a bulk sample representative of the average composition of the lot.

At a specified threshold level (per cent GM), to decrease the relative uncertainty, the size of the sample collected should be increased. In addition, the sample size also needs to increase as the threshold level decreases. This is because the lower the relevant concentration (*i.e.* the allowable threshold) of GM material in a seed lot, the more seeds are required in the collected sample to accurately detect and measure it.

ISTA has developed rules for sampling seed lots that are designed to provide confidence that the submitted or laboratory sample is representative of the lot in most cases. The maximum seed lot size for maize should be 40 tonnes (± 5 per cent), and for canola 10 tonnes (± 5 per cent). There should be documentary evidence that the lot being tested is all from the same source and batch. ISTA rules define the minimum number of primary samples to be collected, in containers ranging from less than 15 kg up to the maximum permitted (and also for a seed stream), and state that the primary samples should be approximately the same size as each other. In the case of a 10 tonne seed lot for any seed, the ISTA rules specify the collection of twenty 500 g primary samples.

The minimum number of seeds that needs to be collected from the bulk sample for analysis depends on the relevant threshold level for AP of GM seed and the level of confidence required in the result. Based on these two factors, a statistical approach can be used to determine the minimum number of seeds required in a sample to ensure, with a defined of confidence, that at least one GM seed is present. This approach is based on two assumptions: the bulk sample is homogenous, and the number of GM seeds present in any working sample taken from such a bulk sample which contains GMseed follows a binomial distribution. These are fairassumptions provided that appropriate protocols (ISTA) are followed for mixing and dividing the bulk and laboratory samples.

For instance, if a homogeneous bulk sample contains 0.9 per cent GM seed, a sample of 332 seeds collected from the bulk sample will contain at least one GM seed 95 per cent of the tim(Table). In other words, if a buyer wants to be 95 per cent confident that seed lots with adventitious GM presence at the level of 0.9 per cent or more in the bulk sample will be detected, then the working sample needs to contain a minimum of 332 seeds. To increase thbuyer's confidence to 99 per cent, the number of seeds analysed would need to increase to 510.

These estimates for the minimum number of seeds to be analysed are valid for use with a *qualitative* analytical method only if the method has no false positives or false negatives. Under these circumstances, a positive result from a qualitative test would indicate, with thdefined level of confidence, that presence of GM material in the bulk sample exceeded the threshold level (in this example presence of GM material greater than or equal to 0.9 per cent).

The measurement uncertainty associated with DNA-based*quantitative* assays is relatively large due to the intrinsic nature of the amplification process of the polymerase chain reactioassay that is used to detect the DNA. Sampling error is not likely to be a significant contributoto the total uncertainty provided that good sampling practice is used—small sample sizes should not be chosen and the sampling error should be below 20-30 per cent. The total measurement uncertainty of a quantitative result should be taken into accounif such an assay is to be used to screen for the presence of GM material in grain and seed.

If the analytical method is capable of detecting a single seed in this minimum number of seeds, it can be used directly on a representative working sample of the appropriate size. However, in some cases the chosen analytical assay may not be sensitive enough to detect a single GM seed in the sample. If this is the case, then the sample should be split evenly into several working samples and the number of seeds in the working samples should be selected to ensure that the analysis is conducted within the detection limit of the assay. Each of these working samples should be analysed using the same method.

ZERO PRESENCE LEVELS

Analytical testing to a zero presence level is not possible. It is possible to test only as low as the limit of detection of the analytical method will allow. For example if a test is capable of detecting down to a concentration of only 0.1 per cent GM (that is one GM seed in 1 000 non-GM seeds), then a negative result (no GM material detected) does not mean that there is no GM material present. There could still be GM material present at a concentration lower than 0.1 per cent. Therefore, detection will always be limited by the sensitivity of the analytical testmethod and the number of seeds analysed in the working sample(s).

SEED COLOUR TESTING AND SAMPLING

Tests of the quality and other characteristics of seed need to be made at several stages during its progress from the parent tree to the seed bed. Tests for maturity and soundness of seed before and during collection in the forest have been described on pp. 34–36 and are usually associated with surveys of the abundance of the seed crop; they are intended to ensure that both the quantity and the quality of seed justify the effort and cost of collection. Several tests may be required at the seed processing depot, after extraction and cleaning and before the seeds are despatched to the nurseries or placed in storage.

Testing of moisture content is essential for many species before they are stored for a long time, but is usually unnecessary if the seed is to be sown immediately in the nursery; on the other hand, partial drying may be needed before despatch of seed to a distant forest station, to reduce the risk of deterioration in transit, and any deliberate drying operation must be accompanied by the testing of moisture content. Testing of germination or viability should be repeated at the end of the storage period if this is more than a few months and, in the case of long-term storage for conservation of genetic resources, testing should be done at intervals throughout the storage period.

Efficiency and success in raising plants in the nursery and in their subsequent establishment in forest plantations depend to a great extent on the quality of the seeds used. It follows that foresters, nurserymen, seed dealers and others need accurate estimates of the quality of the seeds in which they deal or which form the basis of their afforestation projects (Turnbull 1975 d). This is particularly important when seeds are bought and sold, or move between countries in international trade. In practice, the number varies considerably from year to year and from seed lot to seed lot, so that it is necessary for the forester to have an accurate estimate of germination for every individual seed lot which he receives, if he is to fulfil planting targets while at the same time avoiding the waste of valuable seed caused by sowing too much. Similarly the nurseryman must have a good estimate of germination if he is to adjust sowing rates to produce the optimum spacing of seedlings in the seedbed.

Justice has defined the following objectives for developing rules for seed testing:

- To provide methods by which the quality of seed samples can be determined accurately;
- To prescribe methods by which seed analysts working in different laboratories in different countries throughout the world can obtain uniform results;
- To relate the laboratory results, insofar as possible, to planting value,
- To complete the tests within the shortest period of time possible, commensurate with the above-mentioned objectives; and
- To perform the tests in the most economical manner.

For the practising forester, the most important object of seed testing is the provision of an accurate estimate of the capacity of a given seedlot to produce healthy, vigourous plants suitable for field planting. In the present context, seed "quality" refers to the physiological vigour of the seed rather than its genetic quality. The essence of good seed testing is the application of reliable standard methods of examination to ensure that uniform and reproducible results are obtained (Turnbull 1975 d). Standardisation has been greatly facilitated through the adoption by a number of countries of the International Rules for Seed Testing formulated by the International Seed Testing Association (ISTA).

ISTA was founded in 1921, drew up its first set of Rules in 1931 and made major revisions of the Rules in 1953, 1966 and 1976. In the early stages ISTA's primary concern was for agricultural seeds, but trees and shrubs have gradually assumed greater importance and the 1976 Rules (ISTA 1976) give (firm) "prescriptions" or (more tentative) "suggestions" as to testing methods appropriate for 61 different genera of these, compared with 26 in the 1953 Rules. The Rules are published in English, French and German.

Although tree species and genera from the tropics and southern hemisphere (notably eucalypts) have begun to figure in the ISTA lists, the balance is still overwhelmingly weighted towards north temperate species. Thus important species such as Tectona grandis, Pinus patula, P. oocarpa and P. kesiya are listed, but only in the list of "suggested" (non-prescriptive) seed testing methods, while a recent survey has shown that 20 other important tropical species are not included at all; they include Cupressus lusitanica, Gmelina arborea, Cordia alliodoraand the genera Albizzia, Araucaria, Casuarina, Swietenia, Terminalia and Triplochiton. Their omission simply reflects the lack of reliable information from research on the best methods of testing these species and genera.

ISTA testing methods involve the maintenance of controlled laboratory conditions and the use of some items of equipment which are relatively expensive. They are therefore well suited to large, well-equipped seed laboratories but are impracticable for small forest stations which have no laboratory and must carry out germination tests in the nursery or the office. Lack of equipment is no reason to omit seed testing altogether. For example simple germination tests carried out in the nursery can produce results which are perfectly satisfactory for local use. But precise details of method should always accompany results whenever this is not standardised, so that the seed user can interpret the results in relation to his own nursery conditions.

An up-to-date guide to the range of equipment available for seed testing is contained in the "Survey of Equipment and Supplies" published by ISTA (ISTA 1982). It lists equipment by type, use, model specification and supplier. The earlier directory, "Equipment and supplies for collecting, processing, storing and testing forest tree seed" (Bonner 1977) includes seed testing equipment and gives the addresses of user laboratories as well as suppliers. Van der Burg *et al.* (1983) give a description of how a seed testing station in tropical or subtropical areas could be established. Two alternatives are described: Seedlab 2000, that can test about 2000 samples per year and Seedlab 5000 that can test at least 5000 samples per year.

Directives and general considerations concerning the staffing, the organisation of the work, the lay-out of the building, and the equipment needed are given. Forty-six figures and two tables give an impression of the equipment and administrative forms used. Equipment which in the experience of the authors has been found suitable for the work is recommended and detailed descriptions and company addresses are mentioned. Some equipment that is not commercially available is described and plans for construction are included. A list of books and journals for a basic seed testing library is appended. The tests which may be required are purity, authenticity, seed weight, germination, indirect testing of viability, moisture content, and seed health and damage. A prerequisite for all testing is good sampling, which is described in the following section.

SAMPLING

When seed is tested, the sample on which the test is made has to be representative of the whole. No matter how accurate the technical work in the test, the results can only show the quality of the sample submitted for analysis (Aldhous 1972). Therefore, every effort must be made to ensure that the submitted sample accurately reflects the composition of the whole seed lot.

This is equally applicable, whether a whole series of tests is to be made in a laboratory, or a simple nursery test carried out to determine number of germinated seeds per kg of uncleaned seed. Time spent testing carelessly drawn samples may be time wasted (Carter 1961). Completely homogeneous seed lots would be easy to sample, but they do not exist (Bonner 1974). Common sense measures can be taken to reduce heterogeneity as much as possible, for example seed lots of the same species should not be mixed if they come from different origins (provenances), differ greatly in age or, in the case of the same introduced provenances, if they come from plantations growing on very different site types in the introducing country.

But a seed lot may be heterogeneous even if collected from a homogeneous stand. Paul has given the simple example of pine seed transported over bad roads in the back of a landrover. The bumps and vibrations to which it is subjected will cause a settling out of different types of material. All the small and empty seeds and other light rubbish will eventually reach the surface layers and a sample taken from the top alone will give a completely false impression of the potential performance of the seeds as a whole.

MIXING

If the seed lot is a small one consisting of only a few kg of seed, it is possible to improve its homogeneity by thorough mixing of the whole lot before taking a sample. In the case of very large seed lots transported and stored in many different containers, mixing of the whole seed lot is impracticable. Instead, a number of different samples are taken, as described on pp. 194–195, and it is these samples which are thoroughly mixed together to form a homogeneous "composite" sample for testing. The methods of mixing which are quoted below from Paul, are equally applicable to small entire seed lots or to composite samples.

"Methods of mixing.

There are two simple ways in which seed may be thoroughly mixed:

1. *Mixing with the aid of a mechanical divider:* Mechanical dividers are used to reduce the size of seed lots or samples by repetitive halving. Their operation is described below. The divider has the added advantage that it can also be used to mix seed in the following manner:
   - Pass entire seed lot through divider.

- Take the two equal portions and feed them simultaneously into the divider.
- Repeat operation No. 2.
- Repeat operation No. 2 once more.
- Take the two equal portions and pour them simultaneously into the storage container.

2. *Handmixing:* It is surprising how difficult it can be to achieve a homogeneous seed lot using this method. The entire seed lot must be spread out on a sheet of paper or some other suitable smooth surface and mixed by scooping the seed from side to side and from top to bottom. After thoroughly mixing, spread the seed out evenly and divide the lot into four equal parts. Place these separately in four containers and with the aid of an assistant pour them simultaneously into the storage container. Repeat the spreading, quartering and pouring process another two times."

USE OF SEED TRIERS

For a large seed lot contained in a number of different containers, seed triers are used to take small sub samples from different parts of the lot. All of these "primary" samples are mixed together to form a "composite" or bulk sample, which is later reduced in size by successive dividings until it is small enough to form the "working" sample on which the various tests are carried out.

As described by Turnbull (1975 d), a seed trier is a probe, long enough to reach all areas of the seed bag and designed to remove an equal volume of seed from each area through which it travels. Most commonly used is a sleeve-type trier which consists of a hollow brass tube inside a closely fitting outer sleeve. The tube and sleeve have open slots in their walls so that, when the tube is turned or slid, the slots in the tube and the sleeve coincide and seeds can flow into the tube; when the tube is given a half turn or slid in the opposite direction, the openings are closed. Seed triers should have a series of separate compartments and not just a single long tube with several openings. Tubes are available in varying lengths and diameters to cater for different sizes of container and kinds of seeds. Triers known as "thief" triers, with a single open slot, should be avoided, as they can damage the seeds.

Ideally, the primary samples should be distributed proportionally to the volumes of the different parts of the seed lot (Bonner 1974). For example, if a lot is contained in ten 10-kg containers and ten 20-kg containers, the 20-kg containers should contribute two thirds of the composite sample and the 10-kg containers only one third. For seed lots in containers of uniform size, ISTA (1976) has detailed prescriptions as to the number of primary samples appropriate for varying numbers of containers, *e.g.*, for 6 – 30 containers at least one in three containers is to be sampled and never less than five. Large seeds and others that are not free-flowing are not suitable for sampling by seed trier. They should be sampled by thrusting the hand into the seeds and removing small portions (Bonner 1974). The hand should be inserted flat with the fingers extended together. The fingers should be kept together as the hand is closed and withdrawn. It is difficult to sample with this method deeper than about 40 cm and containers may have to be partially emptied to facilitate sampling.

Reducing the Size of Composite Samples

The following description of the reduction of the size of the composite sample is taken largely from Turnbull. Usually the composite or bulk sample for testing needs to be reduced to a working sample of standard weight. Whenever possible the sample should be subdivided by a mechanical divider to reduce any operator bias in the dividing. The non-mechanical halving method is described here for use if a mechanical divider is not available, but it is the least desirable method for most types of seed. Manual methods must, however, be used for seeds which are not free-flowing.

Non-mechanical Methods of Dividing

Halving method. The sample is placed on a clean surface and thoroughly mixed by hand; it is then divided into quarters with a sharp-edged spatula and opposite quarters discarded. The process is repeated until a final sample of the approximate weight required is obtained. Random cup method. A series of small cups or thimbles are arranged on a tray in a definite pattern and the sample is poured systematically over this area. The working sample is obtained from randomly selected containers. A modification of this method uses a tray divided into an equal number of square compartments, every alternate one of which has no bottom.

Mechanical Methods of Dividing

Most mechanical dividers are designed to split the sample into two approximately equal parts. The working sample is obtained by repeatedly dividing the bulk sample until it reaches the required weight. ISTA recommends three types of dividers as being suitable equipment. They are the conical divider (Boerner type), the soil divider, and the centrifugal divider (Gamet type). They are briefly described as follows: Boerner Sampler. This divider is made in different sizes. The essential parts consist of a hopper, inverted cone, and a series of baffles directing the seeds into two spouts. The baffles form alternate channels and spaces of equal width. They are arranged in a circle at their summit and are directed inward and downward, the channels leading to one spout and the spaces to an opposite spout. A valve or gate at the base of the hopper controls the seed flow. When the valve is opened the seeds fall by gravity over the inverted cone where they are evenly distributed to the channels and spaces; then, they pass through the spouts into the seed pans below. In Zimbabwe a similar arrangement on a larger scale is used for dividing and mixing large seed lots. From the inverted cone there are 8 outlets which pass the seed into buckets, and a battery of 8 buckets is available for each outlet. The buckets hold 6 kg of pine seed each, so that a seed lot of up to 384 kg can be handled in any one dividing and remixing operation (Seward 1980).

Soil Divider. A simple divider, built on the same principle as the conical divider, is the so-called soil divider or riffle divider. The channels are here arranged in a straight row instead of a circle as in the conical divider. The soil divider consists of a hopper with attached channels or ducts, a frame to hold the hopper, two receiving pans and a pouring pan. In using the divider the seed is scattered fairly evenly in a pouring pan and poured in at approximately equal rates along the entire length of the hopper. The divider is suitable for large-seeded and chaffy species, but suitable types for small-seeded species can also be made.

Gamet Divider. The Gamet divider makes use of centrifugal force to mix and scatter the seeds over the dividing surface. The seeds flow downward from a hopper onto a shallow rubber cup below. Upon rotation of the rubber cup by an electric motor the seeds are thrown out by centrifugal force and fall downward. The circle or area where the seeds fall is equally divided into two parts by a sharp-edged, stationary baffle so that one-half of the seeds fall in one spout and the other half in the other spout. In using this divider care must be exercised in dividing very small samples as it is possible that a majority of seeds may be thrown out in one spout. Hardin *et al.* (1965) compared the effectiveness of the Boerner and Gamet dividers and concluded that if used correctly, the Gamet divider yields a slightly more accurate subsample.

Magini (1962) warns that dividers operating on the centrifugal principle should not be used for seed sensitive to damage by vibration as are those of Abies species.

Weight of Sample

The weight of the working sample depends on the seed size of the species in question. The ISTA rules are intended to provide a minimum of 2500 seeds for all except the very large-seeded species, for which a minimum of 500 seeds is prescribed. These quantities are considered sufficient for most of the usual tests (purity, authenticity, seed weight, germination or viability), but an additional 10 g of most species is needed if moisture content is to be determined (ISTA 1976). Among tropical species listed by ISTA the range of prescribed minimum size of working sample is from 2 g for Eucalyptus deglupta, which averages 4,000 viable seeds per gm of "seed plus chaff" (Boland *et al.* 1980) to 1 kg forTectona grandis which averages 2,000 fruits per kg. When seed is despatched to an independent seed laboratory for testing, ISTA recommends that the weight of the submitted sample be double that of the working sample.

Purity Analysis

Tree seed samples can contain impurities such as weed seeds, seeds of other tree species, detached seed structures, leaf particles and other material. The object of purity analysis is to determine the composition by weight of the sample being tested. To do this the sample is separated into component parts. When purity analysis is done, it is the first test to be carried out because subsequent tests are made only on the pure seed component. Pure seed refers to the species under consideration and in addition to mature, undamaged seed includes: undersized shrivelled, immature and germinated seeds, provided they can be definitely identified as the species under consideration; and pieces resulting from breakage that are more than one-half their original size (ISTA 1976). Seeds of Leguminosae and Coniferae with the seed coats entirely removed are regarded as inert matter. Other components of the sample may include Other seeds of all species except that under test and Inert matter. Inert matter consists of pieces of broken or damaged seeds less than half the original size, wings of coniferous species, leguminous and coniferous seeds with the seedcoat entirely removed and other matter such as fragments of leaves, twigs, stones, soil. In Coniferae (except Chamaecyparis,Cupressus and Thuja) any remaining seed wings not already removed in cleaning should be detached and classed as inert matter.

The working sample containing all the impurities is weighed and then the pure seed is removed and weighed separately. ISTA rules (1976) prescribe that weighing be done in gms to the minimum number of decimal places necessary to calculate the percentage of its component parts to one decimal place.

The percentage of pure seed is calculated as follows:

$$\text{Purity \%} = \frac{\text{Weight of "Pure" seed}}{\text{Total weight of original sample}} \times 100$$

When a laboratory is testing a large number of samples for purity, the purity analyst usually examines and separates seed samples on a "workboard" placed on top of a desk or table. The workboard can be adjusted to a desired height, usually 7 – 15 cm above table height. The analyst's equipment should permit him to do his work with a minimum of effort and time, and also with a minimum of eye strain (Justice 1972).

Equipment required for making purity analysis is as follows: forceps and spatula for handling, separating, and pushing seeds over a surface; wide-field hand lens of 5X to 7X magnification; reading magnifiers relatively free of curvature and distortion, wide-field stereoscopic microscope with a range of magnifications from 10X to 75X; scale with capacity up to 1000 g and sensitive to 0.5 g; torsion balance with capacity of 120 g and accurate to 0.01 g ; rapid-acting chemical balance accurate to 1 mg; small seed containers to hold separations; and seed blower (Justice 1972). An assortment of small sieves can help in eliminating many impurities.

Analyses of many samples can be facilitated by removing the lightweight material with the aid of a seed blower. Other methods reported by Turnbull (1975 d) as reducing hand labour are specific gravity separation by differences in density, separations by electrostatic charges, a vibrating table working on the basis of seed weight, surface texture and shape, and an X-ray radiograph technique.

The pure seed component from the purity analysis can be subsampled for the germination test, as well as for the determination of seed weight. Since the germination test is based on pure seed, it can be readily seen that the purity analysis and germination test are complementary to each other. The production potential of a seed lot can be determined only when the purity analysis and the germination tests are considered together (Turnbull 1975 d).

SEED WEIGHT

Measurement of seed weight is made on the pure seed component separated by the purity test. Weight is normally expressed as the weight of 1000 pure seeds. This figure can be readily converted to number of pure seeds per g or per kg as required. Weight may be determined simply by counting out 1000 seeds and weighing them (Bonner 1974, Paul 1972), but the use of several smaller samples enables the analyst to estimate variation within the sample. ISTA (1976) prescribes 8 replicates of 100 seeds each, from which the standard deviation and coefficient of variation may be calculated, as well as the mean.

If the co-efficient of variation is less than 4, the mean is accepted, but, if it is more, 8 additional replicates are prescribed, a new standard deviation for the 16 replicates calculated and any replicate that diverges from the mean by more than twice the standard deviation is discarded before calculating the final mean for the sample.

The counting of seeds for weight may be done by hand or with the assistance of counting boards or vacuum or electronic counter. The following description of the essential features of seed counting devices is taken mainly from Magini (1962). Counting boards are useful for relatively large seeds of a regular shape, such as Pinus pinea or some leguminous species. The counter consists of two boards, a stationary top board with perforations and a thin solid board below, which serves as a false bottom.

In operating the counter, a quantity of seed is placed at the corner, the board tilted and moved so that one seed goes into each perforation. The board can then be emptied and the known number of seeds from it weighed. Alternatively, if the seeds are being counted in preparation for a germination test, the counter can be placed on the substrate and the bottom is withdrawn permitting the seed to fall through at a uniform spacing. In some types there is a second set of perforations in the bottom board. By slipping the top section about one centimetre, the seed falls through both boards onto the germination bed. In other types, the counting apparatus is perforated by holes too small to allow the seeds to pass through but suitable to allow a flow of air to suck the seeds into position (Tirén 1948).

In connection with tests on soil trays, a simple counting perforated board can also be employed. The board is first placed over the tray. The seeds are placed in the perforations and the board is lifted leaving the seeds on the tray. Vacuum counters can be used with seeds of various sizes and shapes.

For very small and irregularly shaped seeds some difficulties may occur. They are not suitable for very large seeds, such as many nuts. A vacuum counting device has three essential parts: a vacuum system including pipes; counting heads; a quick release valve. A small house vacuum cleaner with a home-made head may be used, but special power units are generally preferable (1/4 to 1/2 hp electric motors).

Perforated heads, with evenly spaced holes (generally 100) to house the seed, hold the seed by suction. When suction is switched off, the seeds are released for weighing or onto the substrate for germination testing. Counting plates or heads are of a square, rectangular or circular form, according to the shape of the substrate employed. Espacement and diameter of holes should be proportionate to the seed size. Heads can be of plastic, chromium, brass or aluminium.

The face plate should be relatively opaque and of a colour which contrasts with the seeds being counted, thus making it easier to check that all holes are properly filled. A good valve is needed to allow a quick release of seed, as well as to avoid sucking up too large a proportion of the lightest seeds. The risk of picking up an excessive proportion of light, empty seeds in the counted sample may be reduced by inverting the head so that the holes are upwards and then pouring a large sample of seeds over it (Robbins 1982 b). Operation of vacuum counters with plastic heads may be adversely affected by static electricity.

In the modern types of electronic seed counters a vibrating feed bowl moves seeds past an electric eye in single file. The machine can be programmed to count a certain sample size, then cut itself off (Bonner 1974). Reservations have been expressed about the accuracy of vacuum and electronic counters on the grounds that they select a biassed sample (Gordon and Wakeman 1978). Manual counting boards, combined with careful random sampling, are likely to be more accurate, although considerably slower.

The 1000 pure seed weight can be converted to seeds per g or per kg as follows:

$$\text{Number of seed per g} = \frac{1000}{\text{weight in g of 1000 seeds}}$$

or

$$\text{Number of seed per kg} = \frac{1000 \times 1000}{\text{weight in g of 1000 seeds}}$$

If the sample counted is other than 1000 seeds, the appropriate formula is:

$$\text{Number of seed per kg} = \frac{\text{No. of seeds in sample}}{\text{weight of sample in g}}$$

or

$$\text{Number of seed per kg} = \frac{\text{No. of seeds in sample} \times 1000}{\text{weight of sample in g}}$$

Within a given species full seeds have a higher specific gravity and a higher germination rate than seeds of the same size but empty or partially filled. Large seeds weigh more per seed than small ones of the same specific gravity and, because they contain larger food reserves, they are likely to germinate better and produce initially more vigourous seedlings. Goor and Barney (1976) have reported that large seeds of Eucalyptus citriodora had a higher germination rate than medium-sized seeds which in turn had a higher rate than small seeds. Number of pure seeds per unit weight is therefore not in itself a good guide to potential for plant production and must be complemented by germination tests or indirect tests of viability. The effect of seed size on the growth of germinated seedlings of eucalypts usually persists for 8–14 weeks after sowing, thereafter other factors may become more important.

GERMINATION TESTING

Of all the quality measurements of seed lots, none is more important than the potential germination of the seeds. The main aim of a laboratory germination test is to estimate the maximum number of seeds which can germinate in optimum conditions. The use of standardised ideal conditions in the laboratory such as those prescribed by ISTA ensures that results obtained from a given seed lot in one laboratory should be identical with those obtained from any other laboratory in the same or in other countries. A common standard for assessing germinative potential is very valuable for seed moving in international trade. On the other hand, it is clear that results obtained under ideal controlled conditions in the laboratory are not directly applicable in the field nursery, where only limited control over environmental conditions is possible.

Each nurseryman should apply his own correction factor, derived from experience over the years, to convert the germination potential of a seed lot, as determined from laboratory tests, to the actual field germination he can expect under his own local conditions. At the other extreme, a nurseryman may prefer to carry out his own germination test in his own nursery in advance of large scale sowing. The results of the test should be directly applicable to the later sowings of the same seed lot in the same nursery, but would not be applicable to other nurseries. However, there is sometimes insufficient time for germination tests in advance of the main sowing and managers of large operational nurseries may be reluctant to involve themselves in small scale research.

Between the two extremes lies the case of the small silvicultural research station which lacks the laboratory facilities to comply with ISTA prescriptions for testing, but which carries out germination tests in the nursery on bulk seed supplies before distributing them to afforestation projects throughout the country. Here again the local forester or nurseryman must apply his own correction factor to convert actual germination figures from the research nursery to expected germination figures in his large operational nursery.

The following account is largely based on that of Turnbull (1975 d). It is in conformity with ISTA rules (1976), but some of the principles are equally applicable if lack of equipment or trained staff makes it necessary to use simpler methods. Germination is defined as the emergence and development from the seed embryo of those essential structures which are indicative of the seed's capacity to produce a normal plant under favourable conditions (Justice 1972, ISTA 1976). Germination is expressed as the percentage of pure seeds which produces normal seedlings or as the number of seeds germinating per unit weight of the sample.

In the laboratory the environmental conditions, including moisture, temperature, aeration and light, must not only be specific enough to initiate germination but also favourable for the development of the seedlings to a stage where interpretation of normal and abnormal types may be made. With a few exceptions, all germination tests should be made with pure seeds separated by the purity test. The pure seed must be well mixed and counted at random into replicates.

The seeds should then be spaced uniformly on the test substrate. Normally one test consists of 400 seeds in 4 replicates of 100 seeds each, but if 100 seeds overcrowd the test substrate, the replicates may be broken down into a larger number of smaller replicates of 50 or 25 seeds each (Bonner 1974). A general recommendation is to leave 1.5 to 5 times the normal seed width or diameter between seeds, in order to discourage the spread of fungal moulds (Bonner 1974, Justice 1972).

The exceptions are very small-seeded species in which it is impossible (some species of Eucalyptus) or very difficult (Alnus, Betula, Populus,Salix) to separate the seeds from the accompanying inert matter or chaff. In these cases the test is made with the same number of replicates but the replicates are of equal weight rather than an equal number of seeds.

Germination Equipment

The kind of germinator can be selected according to the type and amount of seeds being tested and it is acceptable providing it can give good control of prescribed temperature, moisture, and light conditions.

Germinators range in size from small or portable individual units through cabinets of various sizes and large germination tables of the Jacobsen Copenhagen type, to walk-in germination rooms. The major types recommended by ISTA are as follows:

Jacobsen and Rodewald apparatus. In Europe, a germinator called the Jacobsen apparatus or Copenhagen tank is commonly used. It consists of a bath containing water, the temperature of which may be controlled by thermostats. The seeds are distributed on top of paper and placed on metal or glass strips suspended about 5–7 cm above the bath; paper or cotton wicks lie beneath the substrate and pass through slots into the water below. The moisture content of the substrate may be adjusted by altering the water level, and so increasing or decreasing the distance between the seed bed and the water (Kamra 1968).

High humidity is maintained around the seed either by covering the entire tank with a transparent lid or placing an inverted plastic funnel with a hole in the conical end over each germination pad. The apparatus may be exposed to natural daylight but artificial light is usually desirable. One of the disadvantages of the standard Copenhagen tank is the absence of direct temperature control at the seed bed. In the tropics it tends to overheat (Robbins 1982 b). Improved models have been developed, such as the one by Overaa (1962) which uses hollow stainless-steel strips, through which the water for heating and cooling is circulated. Other disadvantages are that the equipment requires a large amount of floor space for the number of tests accommodated. Also, the handling of covers and substrata appears to be more cumbersome than moving a lightweight tray with paper substrate (Justice 1972).

The Rodewald apparatus consists of a zinc box covered with glass in which the seeds are exposed to direct or diffuse light. The bottom of the apparatus contains water over which a layer of moist sand is placed on a tray. The seed bed consists of unglazed porcelain dishes placed in the moist sand, or the dishes may be placed directly in the water. The temperature of the water is controlled thermostatically and the sand is moistened through wicks reaching into the water bath. A sand substrate is not desirable for species requiring alternating temperatures because of its slow adjustment to a change in temperature.

Germination cabinet. Another common type of apparatus is the closed chamber for germinating seeds in darkness, diffused light, or direct light. This type of germinator often consists of a double walled cabinet, suitably insulated against temperature changes by an air jacket or layer of insulating material. It is equipped with appropriate tray slides to accomodate the type of germination trays preferred by individual laboratories. The modern germination chamber has provision for both heating and cooling. Usually, water is chilled and passed between the walls of the chamber or through cooling pipes placed along the inner walls of the chamber. Either the air or a reservoir of water in the bottom of the chamber may be heated. In such cabinets the temperature can be regulated at any desired setting between approximately 8°C and 40°C (ISTA 1976).

Inexpensive germination cabinets are sometimes constructed locally. Gupta and Kumar (1977) describe the low cost germinators in use in the Seed Testing Laboratory in Dehra Dun, India. The outer wall of the cabinet is constructed of teak, the size 195 × 70 × 40 cm, and the electric controls include a thermostat, hot air blower, a separate air circulator and a time switch to control illumination. Glass wool is used for insulation between the inner and outer walls. Satisfactory performance is obtained at temperatures varying from room temperature up to 45°C (± 1°C).

A germination cabinet should meet as far as possible the following requirements (Oomen and Koppe 1969):

- *Air humidity:* As high as possible but preferably never lower than 90 per cent, to prevent the seed substrate from drying up;
- *Air temperature:* Adjustable between 10°C and 35°C, the temperature all over the working section of the cabinet over a number of days to be uniform in each of the temperature regimes, within plus or minus 1°C;
- *Light:* Uniform illumination of the trays, intensity 750 to 1250 lux at the level of the seed;
- *Air movement:* As low as possible to prevent the seeds from drying up;
- *Fresh air supply:* Small, of the order of one air change per hour, which should be sufficient to remove the carbon dioxide produced by the germinating seeds;
- *Day and night cycling:* A rapid initial temperature drop to be achieved in 30 minutes during the day to night change, reaching the ultimate night temperature in one hour, a similar requirement to hold for the night to day change;
- *Condensation:* Not allowed.

However, when cabinets operate with alternating temperatures it is very difficult to maintain the high humidity and the substrates dry out; this necessitates frequent watering, which adversely affects the germination results and adds to the manhours spent on the test.

There are many modifications of the germination cabinet and for detailed descriptions of some of them see Justice (1972) and Oomen and Koppe. Room germinator. When large numbers of tests are made, whole rooms may be used as germinators, with temperature, humidity and light controlled.

GMO LABORATORY TESTING CAPABILITIES

Three laboratories in each of Australia, North America (USA/Canada) and the EU (nine in total) were surveyed by NMI for this study to give an indication of GM testing capabilities in these countries. The information included in this section indicates only whether any of the laboratories surveyed within these regions has indicated a capability to conduct GM testing methodologies.

Laboratory Testing Capabilities for Screening Methods

Screening techniques are conducted to narrow down the possibilities of which adventitious GM events could be present by identifying groups of events with common elements. The capabilities of the surveyed laboratories to conduct screening techniques are not consistent between Australia, North America and the EU. Not all of the screens recommended in the best practice screening packages described in Chapter 11 are currently able to be conducted within the Australian laboratories surveyed.

The 'screening packages' described in the following chapter are for sampling and testing for AP of GM events at several points along the supply chain. They outline details such as the stage in the supply chain, seed lot details, sampling protocol, analysis details and assay protocol. Depending on the situation and also on market demand, the packages could be used to screen for GMOs approved in Australia and for unapproved GMOs in imported commodities.

Once screening methods have been used to narrow down the possibilities of what adventitious GM events may be present in a lot, either trait-specific (protein) or event-specific (DNA) testing methods can be used to identify which event(s) are present within the seed sample submitted to the testing laboratory.

Laboratory Testing Capabilities for Trait-specific and Event-specific Methods

The laboratory capabilities identified by NMI do not appear to allow for comprehensive GM AP testing solely within Austalia, but this does not mean that these capabilities are not present in other Australian laboratories not surveyed for this study.

Canola

From the 2008-09 season, Australia will be growing both GM and non-GM canola. If required, it may be necessary to test—for the purpose of meeting customer requirements in both domestic and export markets—declaring non-GM canola grain for the AP of Regulator-approved GMPs. Australia also imports canola seed for breeding, and potentially grain from time to time, from countries where GM canolas are commercially grown. As discussed previously in this report, Australian seed breeding companies already carry out comprehensive AP testing in generating commercial seed-for-sowing so additional testing may be required only to confirm the integrity of imported non-GM canola grain.

Based on the laboratories surveyed, it appears that ten of the fifteen canola events of interest cannot be identified within the USA, Canada and the EU (Table). However, it is important to not that in the laboratories surveyed, there is a qualitative and quantitative event-specific test available in Australia for the GM canola lines intended for commercial release in Australia and approved by the Regulator (GT73, MS8 and RF3). Five other GM canola events have been approved for commercial release by the Regulator (T45, Topas 19/2, MS1, RF1 and RF2); and, as mentioned above, there are no intentions to commercialise these lines in Australia at this time. Of the remaining lines in Table, some (GT200 and 18) were never commercialised, and others have not been sold in North America for a number of years (T45, last seed sales in 2005; Topas 19/2, MS1, RF1 and RF2, last seed sales in 2003; Westar-Oxy-235, last seed sales 2001; and, last seed sales 1998).

Lateral flow test strips (trait-specific) may be commercially available for some GM canolas of interest; however, the Australian laboratories surveyed for this report did not report that they had such tests available. This does not mean that such tests are not used by other Australian laboratories not surveyed for this study.

Cottonseed

Australia imports cottonseed for use as germplasm in domestic cotton breeding programmes. However, as discussed earlier, such germplasm would be extensively tested by the seed breeding companies to ensure that it is either non-GM, or if it is intended to be GM, that only the intended GM events are present and that there is no AP or unintended GM events. Traditionally, Australia has not imported bulk shipments of cottonseed for processing into food and feed. As the majority of domestically produced cottonseed comes from GM cotton, if required by the market it may be necessary to test any identified non-GM cottonseed intended for export markets for the AP of Regulator-approved GMOs. There does not appear to be a demand for non-GM cottonseed in the domestic market. Of the laboratories surveyed by NMI, it appears that fourteen of the nineteen cotton events of interest cannot be identified by trait or event-specific methods within Australia, two of which also cannot be identified within the USA, Canada and the EU. However, from Table, it is important to note that in the Australian laboratories surveyed, there are event-specific methods available to test for all the GM cottons except for MON88913 approved by the Regulator for commercial release in Australia. Some other GM cottons may have been used in field trials but have never been commercialised.

Soybean

Australia may import soybean for use as germplasm in breeding programmes and grain for processing (crushing). Soybean seed imported for use as germplasm undergoes extensive testing by seed companies for the AP of GM events. As stated earlier, no GM soybean is approved by the Regulator for commercial release in Australia. Of the laboratories surveyed by NMI, it appears that of the twelve soybean events of interest, ten cannot be identified by trait or event-specific methods in Australia and eight of these events also cannot be identified within the USA, Canada and the EU. This does not mean that the remaining events cannot be tested by other laboratories not surveyed. Further, it is important to note that a number of these events, whilst they may have been reviewed for planting in the USA, have never been commercialised.

ACCREDITATION IN GMO TESTING LABORATORIES

The techniques described in the above section are used within a GMO testing laboratory to achieve harmonisation and standardisation of analytical techniques. The use and effectiveness of these techniques are demonstrated in industry by accredited compliance with the appropriate International Standard. 'ISO/IEC 17025:2005(E). General requirements for the competence of testing and calibration laboratories' (International Organization for Standardization 2005) was initially issued by ISO in 1999 and updated in 2005, and is the main standard used by testing and calibration laboratories.

ISO/IEC 17025:2005(E) is a tool for a testing laboratory to demonstrate their competence on an international scale in both laboratory management and technical capability. Accreditation to ISO/IEC 17025:2005(E) is not a blanket endorsement. Accreditation is only given for specific methods that have been audited and shown to conform to the standard. Accredited methods are reassessed periodically to ensure their continued compliance, and accreditation can be rescinded if the requirements of the standard are not met. If a laboratory is not accredited, it does not mean that it is incapable of doing the testing, but it means that it has not demonstrated its capability to the accrediting body. To maintain accreditation, laboratories are required to participate in relevant proficiency testing programmes.

Testing laboratories demonstrate their compliance with ISO/IEC 17025:2005(E) by gaining accreditation from the relevant National Accreditation Body. In Australia, the National Association of Testing Authorities (NATA) is the only National Accreditation Body, and it is also a member of the International Laboratory Accreditation Cooperation (ILAC). ILAC is an international organisation with the aim of facilitating trade by promotion of acceptance of accredited test and calibration results.

The ILAC Mutual Recognition Arrangement provides a peer-reviewed evaluation that accreditation bodies maintain conformance with their own relevant ISO standard (ISO/IEC 17011), and in so doing ensure that laboratories accredited by them are truly compliant with ISO/IEC 17025:2005(E). A similar function is performed by the Asia Pacific Laboratory Accreditation Cooperation (APLAC), but APLAC membership is restricted to National Accreditation Bodies within the Asia Pacific Region. In the USA there are multiple national accreditation bodies capable of accrediting to ISO/IEC 17025:2005 (E), including the American Association for Lab Accreditation (A2LA), Assured Calibration and Laboratory Accreditation Select Services (ACLASS), the International Accreditation Service, Inc (IAS), the National Voluntary Laboratory Accreditation Programme (NVLAP) and the Laboratory Accreditation Bureau (L-A-B). Canada has the Standards Council of Canada (SCC).

There are many national accreditation bodies in the European Union providing accreditation to ISO/IEC 17025:2005(E) including the United Kingdom Accreditation Service (UKAS) in the United Kingdom, in Germany the Deutches Akkreditierungssystem Prufwesen (DAP), and in Belgium BELAC, the Belgian Accreditation Structure.

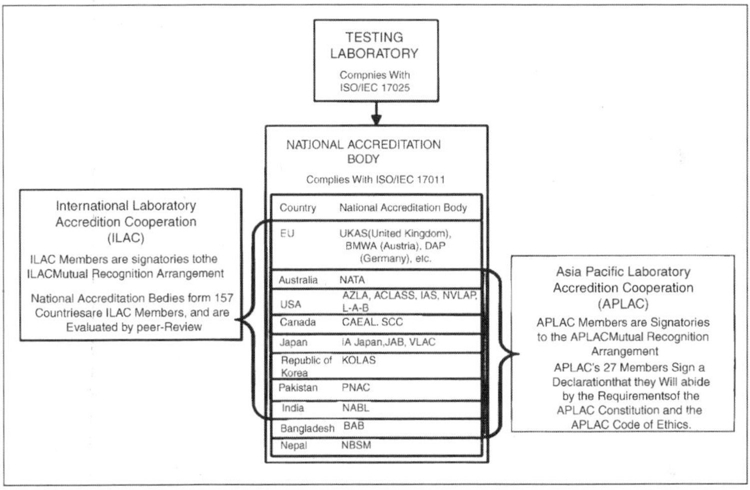

Fig. Laboratory Accreditation Process

Limitations to Use of Sampling and Testing Methods

There are intellectual property restrictions on the use of analytical methods designed within individual laboratories for the detection and/or quantification of GM adventitious presence. Unpublished methods are the property of the laboratory that developed them, whereas standard methods or published methods have no restrictions on their use.

CURRENT SAMPLING AND TESTING FOR GM IN THE SUPPLY CHAIN

Australia currently imports canola seed for breeding purposes, use by seed companies for seed increase, and also as grain for processing for food or feed or for oil extraction. Grain and seed of soybean and maize are also imported, mainly for crushing for processing for food or feed but also small quantities for breeding or for seed increase. Most imported canola and maize is from New Zealand. Cotton germplasm is imported for breeding from time to time. Imported seed of maize, soybean and cotton (but not canola) require quarantine becuase they are 'quarantine restricted species'; restrictions relate to disease and pest risks, not to any risks arising from potential AP of GM events.

Some of the countries from which the seed and grain is imported have either commercialised or are in the process of reviewing for commercial release GM events for these four commodities. No sampling and testing for AP of GM events in imported bulk commodities is required at the border; and no quarantine risk associated with a specific GM event has been identified with any type of import of these commodities. The *Quarantine Act 1908*(Cwlth) requires prior approval via an import permit to import declared GM seeds and grain. AQIS relies on the declaration by the importer as to a shipment's GM status.

In the case of canola seed currently imported for plant breeding (germplasm enhancement), testing for the presence of both Australian-approved and unapproved GMOs is undertaken by industry.

Only GM events approved by the Regulator are allowed to be bred into canola, maize, cotton and cotton breeding lines. Any imports of seed which are GM would require a licence from the Regulator, unless already approved for commercial release in Australia. Conventional non-GM seed imported into Australia by seed companies would be certified seed.

At the time of writing, the only GM broadacre crop in Australia to have been grown commercially and processed through the complete supply chain is GM cotton. GM and non-GM cottonseed was segregated by the Australian cotton industry when GM cotton was first introduced, but following the lack of significant market demand for non-GM cotton, the latter is now delivered in a single supply chain. As a result, there is no requirement for the cotton industry as a whole to segregate or to carry out sampling and testing for AP of GM events past the seed breeding stage. There are few examples of non-GM cotton being segregated.

The Australian Bureau of Agricultural and Resource Economics (ABARE) reported that in 2005 around 20 000 t of Australian cottonseed was certified as derived from non-GM cotton and virtually all of this was exported. For this reason, the cotton industry does not serve as an example or model for GMO sampling and testing needs and capabilities in a coexistence framework, or for managing segregation, for the Australian grain supply chain generally or for the canola supply chain specifically.

An AP threshold does not apply to cotton production in Australia. Furthermore, no GM soybean or maize is grown in Australia. Therefore, the only crop for which there is an AP threshold for GMOs in a non-GM crop is canola. Australian seed companies carry out appropriate sampling and testing of imported non-GM seed for breeding to ensure that it is free from the AP of any GMOs not approved by the Regulator. Sampling and testing is performed according to the ISTA guidelines and their own internal protocols. Such sampling and testing maintains the integrity with respect to the non-GM status of the commercial non-GM seed supplied to growers.

In summary, the only part of supply chains where sampling and testing of conventional non-GM seed and grain for AP of GM material is currently conducted is in breeding and supply of certified seed by breeders and seed companies, as this is a critical stage where unapproved GMOs could enter supply chains.

Importing seed-for-sowing without testing would be a potential source for the introduction of unapproved GMOs into farm production systems, if the seed was imported from a country which grew GM crops and was intended for direct use by farmers in farm production. However, we are unaware of any canola, soybean, maize or cotton seed that is imported into Australia for such direct use by farmers; but if it was, it would need to be certified seed. Seedis imported in small quantities either for breeding or for seed increase by seed companies; or, in the case of soybeans and sometimes maize, in larger quantities for processing for food or feed. All imports of GMOs require authorisation under the *Gene Technology Act 2000* (Cwlth) and any release of a GMO to the environment requires a licence from the Regulator.

SEED AND GRAIN INDUSTRY

The SVGA process resulted in the grains industry agreeing to the*Principles for process management of grain within the Australian supply chain: a guide for industry in an environment where GM and non-GM grain is marketed*. This document underpins the Australian grains industry approach to coexistence, describing the Principles that the industry will adopt in order to maintain their product integrity and meet domestic and export customer requirements in any grains commodity. This document contains the standards, quality assurance practices and sampling and testing regimes which supply chain participants may use to meet customer requirements.

Reference to specific sampling and testing regimes in the SVGA document is currently limited, due to the desire not to impose a specific sampling and testing regime across the entire industry. This approach acknowledges the fact that a sampling and testing regime that works in one sector of the industry may not necessarily be practical in another.

The sampling and testing screening packages for AP of GM events presented in this report provide a level of detail and transparency that may be of benefit to the grains industry, their customers and others. They could also serve as models or be adapted or refined, as appropriate, for use by industry (for example the sampling and testing screening package for canola grain at the grain receival or grain outturn points). Successful development of packages now, when there is no significant GM canola production in

Australi to warrant extensive testing of non-GM canola grain, could position industry for a potential future need, should markets require ongoing segregation.

The need to work towards harmonisation of sampling and testing protocols for GM events was also raised earlier in this report. Harmonisation stems from the need to judge that results are reliable and comparable. Appropriate reference materials are required in order to verify that testing methodologies used in different laboratories are able to produce accurate and comparable results. For example, in an ideal harmonised system, the sampling and testing carried out by an Australian bulk handling company at its grain receival points should be comparable with, and use the same testing methodologies as, the sampling and testing carried out on the grain received at port by an importing country. This would help ensure consistency in results when testing the same seed lot.

In addition to establishing consistency through harmo-nisation, sampling and testing practices will need to be adapted as future GM crops enter the marketplace. The points in the supply chain where seed and grain should be tested for AP may need to be reassessed in a given commodity, and procedures for sampling may also need to be changed accordingly, with changed characteristics of the seed and grain.

TESTING CAPABILITIES

Three laboratories in each of Australia, North America and the EU were surveyed for this study to give an indication of the GM testing capabilities in these countries. Australian laboratory testing capabilities identified in this report include event-specific testing capacity for all GM cotton and GM canola varieties approved by the Regulator for commercial release in Australia. While capabilities do not appear to include capacity to test for all GM events approved overseas, capabilities may exist in overseas and Australian laboratories not surveyed for this study.

The difference in cost for GMO testing between Australia, the USA and the EU is significant. For example, the cost of a qualitative PCR assay in the USA can be as little as US$12.75-209 (AUD$19.03–312), whereas in Australia and the EU it could cost between $400-680 and 109–222€ (AUD$209–426) respectively. This cost differential in testing for AP of GM events is a major reason why Australian seed breeding companies currently send their seed samples for testing to accredited USA and European laboratories. It is also a reason why the bulk handling companies look towards using the lateral flow test strips as opposed to qualitative PCR testing. At about AUD$6–7 per strip test, using lateral flow test strips is significantly cheaper than engaging an accredited Australian laboratory to perform the testing.

Event-specific PCR primer sequences for the GM events are often held as Commercial-in-Confidence by the technology providers and only made accessible to a limited number of accredited laboratories around the world. This could explain why industry may not be able to undertake much qualitative PCR testing in-house. If the cost of testing for AP of GM events in accredited Australian laboratories could be reduced, this could lower overall costs to Australian industry of maintaining product integrity in a GM/non-GM canola coexistence framework, should such segregation be required in the market.

In issuing a licence for dealings involving an intentional release of a GMO into the environment (DIR licence), the Regulator can impose conditions. For DIR licences issued to date, the Regulator has required that 'the licence holder must provide a written instrument to the Regulator describing an experimental method that is capable of reliably detecting the presence of the GMOs and the presence of the genetic modifications described in this licence in a recipient organism.' This testing methodology constitutes intellectual property of the licence holder and as such cannot be shared with commercial testing laboratories.

The International Experience

Approaches to the coexistence of approved GM and non-GM crops in different countries essentially fall into three groups. Some countries, for example the USA, Canada and Australia have adopted a non-legislative approach to coexistence. A second group, such as some EU countries, take a legislative approach (Chapter), and a third group comprises the many countries that are still in the process of considering, developing, or formalising their approach to coexistence.

Coexistence experiences overseas could be instructive aabout how to address sampling and testing for AP of GM events in non-GM seed and grain. In this study, we therefore looked for sampling and testing components in existing and proposed coexistence strategies in the USA, Canada, EU27 countries and the countries to which Australia exports significant volumes of canola and cottonseed.

With the exception of the EU, we found few sampling and testing components that formed part of legislation-based strategies. As highlighted in Chapter, strategies often focus mostly on managing physical admixture through handling methods and gene flow at the on-farm level though physical isolation (or separation) distances between GM and non-GM crops; sampling and testing for GM events is not a common requirement. Whether the coexistence approach is legislative-based or non legislative-based, the specific sampling and testing protocols and methods to be employed are not prescribed, consistent with the approach by, for example, ISTA. However, principles and/or guidance may be provided and international standards referenced, and, if in a legislated framework context, there may be a recommendation or a statutory requirement to comply with them.

Contrasting approaches are exemplified by the following two documents. First is the EC Recommendation on technical guidance for sampling and detection of GMOs and material produced from GMOs in seed, food and feed products (2004/787/EC), which fulfils requirements set out in Regulation No. 1830/2003 on the traceability and labelling of GM organisms and the traceability of food and feed products produced from GMOs. Second is the SVGA Principles—and associated technical documents, particularly the ASF Best Practice Guidelines.

The first document illustrates an approach that is statute-based, with detailed reference in a guidance document to strategies, standards and processes relevant to sampling and testing. The latter documents illustrate an industry-based self-regulatory approach, which also includes detailed reference to strategies, standards and processes. In both these examples, the guidance/recommendations are detailed. The essential difference in approach is whether or not there is associated government intervention through legislation. The industry self-regulated approach relies on breeding programmes and seed companies to have in place quality management systems to ensure compliance and so meet any market demand for segregated product.

The international experience with coexistence models and sampling and testing for GM events can help inform Australia's needs, but does not provide any 'off-the-shelf' approach that can or should be adopted. This is because approaches to coexistence are determined and constructed around two drivers—market-specified demands and statutory regulatory requirements; these vary by commodity and by country, and so sampling and testing needs can become commodity-and country-specific. The sampling and testing needed is that which delivers the commodity which meets the market-specified and/or legally-specified requirements.

In the case of coexisting approved GM crops (approved both for commercial field release and for food) and non-GM crops, the sampling and testing issue is not one of ensuring that some safety threshold has not been exceeded or that an unapproved GMO(s) is absent. One view is, therefore, that the market should determine whether or not there is a need for a coexistence strategy, industry should be responsible for any strategy, and AP thresholds themselves should be what are realistically practical. AP thresholds for approved GM events in non-GM commodity tend to be set at about 0.9 per cent for grain or higher, across different countries. This is essentially the approach that applies in Australia in the case of canola, although Tasmania has effectively set a zero presence threshold, while acknowledging it is not possibleto sample and test to zero presence.

The sampling and testing packages presented in this report are examples of the kinds of best practice packages that can be constructed. However, packages need to be tailored to the specific combination of the following: the

AP threshold to be met; the target confidence levels for the test accuracy, representativeness of the sample, and analysis; the seed lot size; and, the particular commodity (taking into account grain size and weight).

In the case of the sampling and testing that is needed to determine the absence of unapproved GMOs or the absence of any GMOs, the AP threshold is presence. However, analytical testing to zero is not possible, because methods are limited by their sensitivity. Sampling and testing packages for any given commodity therefore need to acknowledge the limits of detection of the methods to be used and to specify the number of seeds to be sampled in a test working sample. A realistic testing plan would specify at least 99 per cent confidence that any AP is below 0.1 per cent.

As described in Chapters, both non-government (ISTA and ISO) and government (for example, the USDA GIPSA and the European JRC and CEN) standard setting bodies already produce comprehensive relevant rules and standards and accredit and certify laboratories.While sampling theory and methodology are well established, testing methodology continues to evolve. More efficient, effective and economic testing methodology is continually being developed. Any organisation or laboratory involved in testing therefore needs tobe aware of current testing methods and select the most appropriate test to meet the purpose required.

POTENTIAL FUTURE SAMPLING AND TESTING

Several points for sampling and testing in current supply chains have been identified which seem practicable (discussed in the previous section). In the case of canola, cotton, soybean, and maize seed imported into Australia as new germplasm for breeding purposes, the key point is at the early seed breeding stage and also, at least, at the breeder seed/field trial 'testing gateway'. Sampling and testing at these points eliminates the prospect that unapproved GMOs can enter domestic production streams for non-GM commodities or, therefore, exported commodities or domestically produced food streams, from varieties or lines bred in Australia.

In the case of imported shipments of these commodities for seed-for-sowing, sampling and testing of seed at some stage before it was sown, either in the originating country and/or in Australia, would confirm the non-GM status of the seed (or, in the case of GM canola varieties approved for commercial release in Australia, confirm the GM variety) and this would further confirm that GM events not approved by the Regulator are absent. A risk-based national strategy to manage the unintended presence of unapproved GMOs in imported seed-for-sowing is in place.

Because a threshold for the AP of a GMO in a non-GMO commodity in domestic supply chains has been set only for canola, sampling and testing for AP is relevant to non-GM canolagrain at bulk handling and/or grain supply stages, including supply for domestic food production and for export, to meet any non-GM market demand. In the case of cotton, although there is no domestic demand for a non-GM supply and therefore no need to sample and test for AP of GMOs; there could be a need for sampling and testing in the case where a niche overseas market required non-GM cottonseed.

In practical terms, the canola industry has indicated that bulk handling companies could monitor non-GM canola for the presence of GM grain at grain receival should markets require segregated product, as sampling and testing for other characteristics is routinely carried out at this stage. The initial step would be to determine the presence or absence of any GM grain in grain to be marketed as non-GM so that it can be appropriately unloaded at the receival point—this could involve a set of industry-designed protocols including declarations and/or testing. Segregations could then be tested at various points as determined by industry protocols and then tested again at grain outturn to meet customer requirements. Depending on these requirements, testing at some points of the supply chain could variously require sophisticated tests (including if there is a need to quantify GMO presence), or perhaps lateral flow test strips may be sufficient in some cases.

It is important to note that GM canola grain would be classed as Standard 'Canola' CSO1, and thus it would not be necessary to test every load of grain at receival points: only those that were to be segregated as CS01-a Standard ('Non-GM Canola').

MODEL SAMPLING AND TESTING SCREENING PACKAGES

The matrix of unapproved and approved GMOs in the four commodities under considerationis complex, and becoming more complicated as more GM varieties are introduced into commercial production. It is complicated because there may be approvals in the country from which we receive imports of seed and/or grain, but not in Australia—and even though approvals for commercial cropping may have been given in a country, the GM crop may never have actually been commercially grown. Further, approvals may be for food in Australia, but not for field release (commercial or trial), or there may be an AP threshold overseas but not in Australia (and *vice versa*).

In this report we have analysed: the current GM events which exist in canola, cotton, soybean and maize varieties in North America; their approval status in Australia (for food and for commercial field release); the production supply chains in Australia; thresholds in place in our major export markets; and, sampling and testing methods and technology. In this context, we have developed model sampling and testing screening packages for potential useacross the four crops at the identified sampling and testing points discussed above.

Whilst the specific screening packages provide feasible options for screening in the current context, it is important to note that they may not reflect what is needed and/or applied in the commercial or regulatory environment. Their currency will depend on which new GM varieties are introduced overseas or in Australia, and at what rate. The screening packages presented in this report could, however, be used as a model approach to ensure continued product integrity in the seed and grain supply chain. They also highlight the potential complexity of sampling and testing for specific purposes.

The sampling process described in Chapter 4 involves taking a number of primary samples from a seed lot, combining these to form a composite sample and reducing it (if it is too large) into a submitted sample to be sent to the laboratory. Key factors associated with this process are the confidence level and relative uncertainty. Understanding the confidence one can have that the results are correct and how far away the actual value could be from the calculated results plays a fundamental role when deciding a sample size. In regard to testing, qualitative PCR tests are the most reliable detection tests and so are to be recommended for determining presence/absence.

The sampling and testing screening packages below have taken the above factors into account when suggesting sampling and testing protocols, target confidence levels and the relative uncertainty (specified in each case).

Depending on the testing results, different decisions will need to be made. For example, if no GM events are detected in a certified non-GM seed or grain lot, it could continue to the next stage in the supply chain with no further sampling and testing requirements. For seed or grain lots, if the screen suggests that a GM event may be adventitiously present in a non-GM lot, further testing could be conducted to determine if the GMO(s) present is Regulator-approved and, if so, the level of GM presence (*i.e.* is it above or below the permitted AP threshold). Such further testing would require the availability of event-specific methods as discussed previously. The outcomes would determine subsequent actions: forexample the lot may no longer be classified as non-GM.

4

Developments of Plant Seed and Spore

The first inception of the living plant is from seed or spore. Subsequent developments based on the physiological processes of photosynthesis and respiration give rise to the mature individual. The birth of the new plant begins with the germination -of the seed or spore. The seed of the higher plants is a highly specialised structure consisting in the majority of cases of a resting embryo with first leaves or cotyledons, plumule and radicle and supplied with reserve food material upon which it draws in the preliminary stages of germination.

The spore of the lower plants, on the other hand, is no such specialised structure as the seed, and is incapable of withstanding prolonged periods of rest before the advent of suitable conditions for germination. The seed with its resting embryo can survive far greater environmental adversities than the less protected spore and is capable of germination on the arrival of suitable conditions.

The necessary factors for the germination of seeds can be classified under four main heads:

- Water.
- Oxygen.
- Temperature.
- Light.

Water is an absolute necessity for successful germination. The quantity of water taken up by different seeds is variable; wheat varies between 40 to 60 per cent., peas from 84 to 106 per cent., and in other plants such as maize, the value lies between 35 and 40 per cent, of the dry weight. The first stages of water absorption are due to imbibition. The cells of the embryo and endosperm are non-vacuolate, and it is only in the later stages of the process that cell vacuoles are formed and osmotic pressure comes into play.

It is still a matter of considerable controversy whether the soaking of seeds in water improves their germinating power. Morinaga has studied the capacity which seeds have of germinating under water in seventy-eight different genera, and has found that forty-three of these genera germinated under water. Seeds of two of the genera germinated better under water than on moist filter paper, while in eighteen of the cases germination under water was equal to that on damp filter paper. On the whole, it was discovered that small seeds germinated better under water than on filter paper. If the oxygen-content of the water were increased, several seeds, normally incapable of germination under water, germinated readily. Temperature also plays a fundamental part in germination. If the temperature be too low, germination fails to occur, if too high, the seed may be permanently injured and killed.

Temperature affects the rate of entry of water into the seed as well as the growth rate of radicle and plumule, it also affects the resistance of the seed coat to the extrusion of the radicle. The temperatures at which different seeds germinate are different. *Phaseolus vulgaris* will show evidence of germination at 9°C., and the growth rate increases with temperature to 36°C. and ceases when 46°C. is reached. Rye will germinate at as low a temperature as 1°C., the maximum being 36°C. and the optimum 25°C. In all cases we are dealing with a minimum, maximum and optimum degree of temperature in germination; although, as F. F. Black-man has shown, the "optimum "points are largely fictitious, varying with the time during which the factor has operated.

The time-factor is of importance here, for the temperature may be too high for further growth to take place, onee the radicle has made its appearance; in other words, for a brief period of time the high temperature has caused rapid growth, but the rate falls away owing to the operation of the time-factor. Morinaga submitted the seeds of a number of genera to alternation of high and low temperature. As a general rule, each exposure to high temperature was for six hours, followed by exposure to a lower temperature for eighteen hours. Thus *Cynodon dactylon* germinated most readily with six-hour periods at 38°C., alternated with eighteen hour periods at 10°C., while alternation between 32°C. and 10°C. and 38°C. and 15°C. were less effective.

If oxygen be excluded, germination is inhibited, while certain seeds fail to germinate in the dark, and light is a requisite factor. According to Morinaga the seeds of *Typha latifolia* germinate more readily when the oxygen pressure of the air is reduced by partial replacement by hydrogen or nitrogen. The action of oxygen and light brings this discussion to the large and important question of the *dormancy of seeds.*

COMPOSED OF AN EMBRYO

It has already been stated that the seed of the higher plant is composed of an embryo which is differentiated into stem or plumule, radicle and cotyledons. The seed, as such, contains a minimum quantity of water, shows an absence of cell vacuoles; and therefore in this condition chemical reactions, especially respiration, are reduced to a minimum. In such a state, the seed can retain its vitality over a prolonged period of time; as much as fifty or eighty years in some exceptional cases. The seeds with the greatest vitality occur among the Leguminosae. As a general rule, seeds which show a high vitality possess a tough testa which prevents the absorption of water and is impervious to the action of gases. The whole question of the factors affecting dormancy of seeds has been reviewed at length by Crocker, and an excellent summary of the physiology of dormancy is given in his paper. In general terms, Crocker found that dormancy or delayed germination can be attributed to one or more of the following causes:

- Incomplete development of the embryo.
- Impermeability of the testa to water.
- Mechanical restraint offered to the expansion of the embryo and other seed contents by the seed coats.
- Inhibition and retardation of gases to, or, from the embryo, resulting in an accumulation of carbon dioxide within the tissues of the embryo, or in an insufficient supply of oxygen for germina-tion.
- The necessity for the embryo itself, to undergo certain after ripening processes before germination and growth under ordinary conditions becomes possible.
- Introduction by various means of a condition of dormancy in seeds previously capable of immediate germination.

To Crocker applied the special name "secondary dormancy." According to Rose, freshly harvested seeds of *Tilia americana* with 10 per cent, or less of water-content, or seeds kept in warm storage for several months, fail to germinate when placed on a moist substratum at ordinary room temperatures. By numerous experiments he was able to show that the germination of the seeds could be brought about by a period of after-ripening in moist storage at 0°C. to 2°C., followed by a period of two or three weeks at 10°C. to 12°C. until germination was well started.

ACTION OF LIGHT ON GERMINATION

The action of light on germination has been much to the fore in recent years, and most of the work on this particular aspect of germination has been conducted in Germany. It has been known for a considerable time that the seeds of some plants do not germinate in the presence of light, while others refuse to germinate in the dark. Busse found that the seeds of tobacco germinated more readily in the light than in darkness. Darkness, however, did not inhibit germination completely, but merely retarded it.

Kinzel, working with the seeds of *Aquilegia atrata,* kept them on moist blotting paper for ten years. During the first year only 10 per cent, of the seeds germinated; the remaining 90 per cent, lay dormant in a perfectly healthy condition for the whole period of ten years until exposed to light, germinated in fifty days. Lehmann showed that the fully imbibed seeds of *Lyihrum salicaria,* of which only a fraction germinated in the dark, gave a germination of about 50 per cent, in twenty-four hours after a brief exposure (1/10 second) to a light intensity of 730 candle-power at a distance of 1 metre. Lehmann supposed that light has some kind of catalytic effect on seeds that will only germinate in the light, converting the insoluble reserve foods into soluble substances. In the case of "dark "seeds, *i.e.*, those seeds that will only germinate in the dark, the light was supposed to activate fluorescent organic compounds which have a destructive effect upon the proteolytic enzymes present.

Further work on *Lythrum salicaria* has been carried out by Lakshmana Rao, who allowed the seeds to germinate under constant conditions of temperature and varying light intensity. At higher temperatures (31°C. to 35°C.) with a single period of continuous lighting the product law was found to apply with very fair accuracy, that is to say, the percentage of germination was proportional to the intensity of the light and the time.

At lower temperatures (25°C. to 28°C.) the action of the light was not so effective, and here germination was inversely proportional to the intensity of the light. These variations from the product law are explained on the results of experiments with intermittent light. In this case the effect was proportional to the sum of the exposures used. With long periods of darkness, the effect of light exposure increased in proportion to the quantity of light used up to a maximum value, while with still longer dark periods, the period of light became less effective again, until finally the effect was less than with continuous light.

Rao concluded from the results of his experiments that light had a direct effect on the embryo and not on the dead tissues of the testa. Many organic and inorganic substances have the power of completely or partially counteracting the retarding influence of light on light-sensitive seeds. Kuhn, working with different seeds, studied the inter-relation of various mineral acids and light and darkness as factors influencing germination. Seeds of *Phacelia tanacetifolia* on filter paper moistened with distilled water gave a germination of over 80 per cent, in the dark, but only 18 per cent, when exposed to light.

By the substitution of weak solutions of different acids, Kuhn was able to raise the germination to 60 to 80 per cent, of that in the light. In the dark, however, the acids appeared to be without effect upon germination. Similar results were obtained with Solarium lyco-persicum and Amaranthus atropurpureus.

Magnus failed to find this encouraging action of dilute acids on the germination of P. tanacetifolia. A so-called "false germination "takes place by the extrusion of the embryo. He also found that an aqueous extract of the seeds tended to inhibit germination in the light, but, curiously enough, not in the dark.

Further, if a portion of the seed near the micropyle were removed, good germination occurred in the light. It is thus evident that some substance is secreted near this portion of the seed that is responsible for the failure to germinate in the light. The action of light on seeds is a complex one. Lehmann found that the seeds of *Veronica tournefortii,* which normally germinate in the dark, at a temperature of 30°C., did not germinate in light or darkness, while solutions of potassium nitrate stimulated germination in either light or darkness; the results, taken as a whole, were better in the dark. Treatment with potassium nitrate in the presence of light resulted in an 80 to 90 per cent, increase in germination as compared with the controls in distilled water.

It was also found that the higher the intensity of the light, the greater the retarding effect upon germination, and potassium nitrate was quite incapable of counteracting this retarding influence. Thus with a light intensity of 440 candle-power at a distance of 1 metre and a temperature of 30°C., potassium nitrate had no effect upon germination. According to Niethammer, with increasing age the sensitivity of the seed to light decreases. It would appear that the place in which the seed ripens plays no part in its sensitivity to light, and he concluded that under natural conditions sensitivity of seeds to light is an important factor in germination. It has usually been supposed that the seeds of *Viscum album* (Mistletoe) require a prolonged resting period of five months before they are capable of germination.

Heinricher was able by suitable experimental conditions of high relative humidity and light intensity of 1,600 candle-power to force the germination of the seeds at any time during the winter. Boerher, in the course of an ecological survey of various forests, has come to the conclusion that light plays a distinct part in the problem of seed germination under natural conditions. In his opinion it has an indirect effect upon soil moisture, evaporation, and other factors. The question of dormancy in relation to age, vitality and respiration of seeds and their oxidase and catalase content has been investigated by Crocker and Harrison.

They were able to show that when the freshly harvested seeds of Johnson grass received a preliminary incubation at 20°C. in a germinator they remained dormant for a year or more. If a portion of the grain surrounding the embryo were removed, prompt germination occurred. It is evident then that dormancy here was brought about by the structures surrounding the embryo. The catalase activity showed a marked reduction while the seeds were in the incubator at 20°C., a reduction of as much as 50 per cent, taking place in the first month.

This fall in catalase content was accompanied by a fall in respiration; whereas there is normally a rise in such seeds kept under conditions of dry-storage. If this gradual fall in respiration be of general occurrence in fully imbibed but dormant seeds, and, further, if the assumption be made that the death of such seeds is dependent upon the utilization and destruction of the reserve foods present by respiratory activity, this result must have an important bearing on the longevity of seeds under natural conditions. Brenchley found that many seeds would survive burial in the soil for as long as fifty-eight years.

LARGE MAJORITY OF CASES TEMPERATURE, OXYGEN, WATER-CONTENT

Although in the large majority of cases temperature, oxygen, water-content and light or darkness are the principal factors concerned in germination, in a number of families of flowering plants germination is found to be dependent upon a fifth factor: the presence of a fungus. This state of affairs is especially prominent in the Orchidaceae and Ericaceae. Unless the seeds of an orchid come into contact with its suitable fungus, no or very little germination occurs. It was on this account, although the scientific reason for the procedure was not understood, that orchid growers always sowed the seeds in soil in which orchids had already been growing.

This association of fungus with higher plant, which is of wide occurrence in the vegetable world, was termed "mycorrhiza" by Frank. The mycorrhizal association of orchids has been widely studied, and now finds commercial application in horticulture. An excellent review of the whole of this subject has recently been published by Rayner, and only certain aspects will be considered here. The classical work of the Frenchman, Bernard, has largely cleared away the difficulties surrounding the germination of orchid seeds.

The seeds of orchids are minute dust-like particles, produced in enormous quantities from a single capsule, and each seed possesses an undifferentiated embryo. The seeds prove to be quite sterile when sown in the ordinary way, but if the necessary fungus, here a species of Rhizoctonia, be present, germination proceeds rapidly and smoothly. In the case of Bletilla hyacinthia, Bernard found that in the absence of its fungus, germination was able to proceed to the first leaf stage, but this, as Ramsbottom pointed out, was probably a "last despairing effort "on the part of the seedling to survive in the absence of its fungus by the formation of its photosynthetic mechanism.

In after-life the fungus is kept within the outer cells of the cortex of the root and never enters the endodermis or vascular cylinder. An investigation of the *role* played by the fungus in the germination of orchid seeds has been carried out by Knudson, who has shown that the germination of the seeds of a hybrid orchid, a cross between a *Lcelia* and *Cattleya,* took place in the presence of sucrose solution and absence of fungus. In the Ericaceae the problem of the relation between plant and fungus has been exhaustively studied by Rayner. Here, unlike the Orchidacea, the fungal mycelium extends through the whole plant and infection of the seeds is brought about in the ovary.

By sterilising seeds of *Calluna vulgaris* in weak mercuric chloride solution, Rayner showed that the root system failed to develop in the young seedling, and that such leaves as were formed soon showed marked chlorosis. Infected seeds, on the other hand, developed an extensive root system and grew normally. This work of Rayner has been called in question by Christoph, who considered that the presence of the fungus was not necessary for the complete development of *Calluna,* and that growth proceeded as readily in sterilised as in unsterilised cultures. The fungus concerned here is a *Phoma,* a member of the fungi imperfccti. According to Rayner, Vaccinium presents a unique case among the Ericaceae and exhibits the most highly evolved symbiotic relation discovered in this family. The fungus, as in Calluna, is distributed throughout the plant and extends to the ovaries and other organs of the flower. Rayner found it impossible to obtain seeds of Vaccinium free from fungus infection by sterilisation or removal of the testas previous to germination.

All the tissues of the emerging seedling were subject to invasion by fungal mycelium, and it was probable that the normal germination of the seeds was bound up with this invasion. The physiological relationship of fungus and higher plant is still a matter of controversy.

It is probable that the mycorrhizal habit originated in a parasitic attack of the fungus upon the higher plant, and that the attack was at first repelled and later turned to advantage by controlling measures, with the final result that a fine balance of adjustment has been attained between the two organisms; the fungus relieving the higher plant of certain of its metabolic functions. Knudson held that the presence of the fungus facilitated the solution of insoluble carbohydrates until the assimilation mechanism had been firmly established.

He found that Rhizoc-tonia repens induced germination of Cattleya seeds on a starch-containing medium and accelerated the germination of orchid seeds in a nutrient solution of sucrose. The fungus was able to effect germination without infecting the seed, due to its ability of digesting starch in the medium and modifying the *pVL* to a value favourable for growth.

PEAT AND SPHAGNUM SUBSTRATE OF NUTRIENT SOLUTIONS

On a peat and sphagnum substrate of nutrient solutions adjusted to a pll of 4-6 germination was as rapid as when the fungus was present; the especial factor for germination being that carbohydrate should be present. In certain cases Rayner found that the fungus in Calluna could play the part of a parasite, especially if the seedling should happen to be at all weakly.

The higher plant must therefore maintain a strict control over its fungal partner to prevent the latter from obtaining the upper hand and destroying the metabolic equilibrium. The metabolic relations in after-life of fungus and higher plant is also obscure. The obvious statement to make in the circumstances is that the fungus must assimilate free nitrogen. Experiments, however, have only to be made on this subject to be immediately contradicted.

More recently, Fulmer and Christensen have grown yeast on a synthetic medium of phosphates and molasses, and found that this plant was able to assimilate free nitrogen at 30°C; the assimilation was said to be a function of the *pH* of the medium. No gain was registered for some six to eight weeks, and it was on this account that Fulmer and Christensen considered that many of the older workers failed to obtain positive results of nitrogen fixation with fungi. Nielson Jones and Llewellyn Smith have brought forward considerable evidence to show that Phoma radicis, the endophyte of Calluna vulgaris, does possess the power of fixing molecular nitrogen. By means of an apparatus entirely closed to the external air, errors due to the presence of ammonia and other gaseous nitrogen compounds were excluded, and the gaseous nitrogen used for the experiments was obtained electrolytically.

The amount of nitrogen fixed was of the order of 0-00095 gm. for 0-345 gm. of glucose used; a value very considerably less than for the nitrogen fixing bacteria. In the absence of either combined or atmospheric nitrogen, the nitrogen value was 0-00016 gm. for 0-21 gm. of glucose used. Other workers have assumed that the fungus in some way assists in the absorption of salts and water from the soil. Exactly how this may occur is difficult to see. McLeimen, from her studies on the endophytic mycorrhiza of Lolium subulatum, L. perenne, L. multiflorum and L. temulentum, found that there was a metabolic exchange between fungus and higher plant, with the result that the higher plant obtained a supply of fat or oil.

Rather than a supply of nitrogenous food material. It is very probable that the metabolic functions of the fungus may be different in different families, and that in one case it may be concerned with the fixation of atmospheric nitrogen and in another with the production and exchange of fats.

VIGOUROUS GROWTH OF THE SEEDLINGS

To obtain vigourous growth of the seedlings a still higher temperature was needed. During the after-ripening process, it was discovered that the hydrogen-ion concentration of the seeds increased as well as the catalase and oxidase activity. Rose also found that the freshly harvested seeds of *Sambucus canadensis* failed to germinate when sown at room temperature, but if they were kept in moist soil out of doors for the length of the winter, a high percentage germinated. He was unable to discover the factors influencing and controlling germination here.

In *Rubus Idceus* he was able to show that dormancy was due to the high breaking strength of the endocarp. The presence or absence of light did not influence the germination of the seeds. Pack ascertained that the seeds of *Juniperus* germinated readily if they were first maintained at a constant temperature of 5°C. for 100 days, to allow of after-ripening of the embryo. If the seeds, after the completion of the ripening process, were removed to a higher temperature (in this ease 15°) they were thrown into a state of secondary dormancy.

The following changes accompanied the after-ripening process:

- An increase in the *pH* of the embryo;
- Increase of titratable acid;
- Decrease in stored fats and proteins;

- Increase in sugar-content;
- Translocation of fats from endosperm to embryo;
- Slight increase in the rate of respiration of the embryo;
- Increase in the value of the respiratory coefficient;
- Marked increase in catalase activity.

Kidd and West have shown that the seeds of *Brassica alba* sown in the presence of carbon dioxide could in certain circumstances be completely inhibited from germinating, and that this inhibition of germination might be maintained over prolonged periods; as much as twelve months in their experiments, during the whole of which time the seeds lay on a moist bed of sand.

When the seeds were exposed to air, Kidd and West discovered that this condition of secondary dormancy could in most cases be terminated by drying the seeds or removing the testas without drying. No evidence of any changes in the seed coat could be found during this period of primary inhibition. On the other hand, it was shown that if the testas were removed with extreme care, the naked embryo remained dormant, and it was therefore suggested that a more stable condition of the tissues of the embryo had become established during this period of primary inhibition. Under the influence of carbon dioxide, embryos in this stable condition did not respond to normal environmental factors which govern the course of germination of ordinary embryos.

Some very definite stimulus, either chemical or mechanical, was necessary to initiate cell division, and therefore the growth of the dormant embryo. The condition of dormancy in the marine pea, *Lathyrus mari-timus,* has been investigated by Stiles and Bellow. The seed of this plant possesses a particularly tough testa which will not absorb water, even after immersion for a year, nor will the seed germinate when planted out. If, however, the testa be broken, germination proceeds normally. Stiles and Bellow suggested that in its normal surroundings, L. maritimus propagates itself either by vegetable suckers or that the seeds are cast on the shingle and the testas cracked by the action of the waves on the pebbles.

It is a well-known fact that willow (Salix) seeds cannot retain their vitality for more than a few days. Yozo Nakajima has shown that the seeds of Salix pierotii and S. japonica lose their vitality under normal conditions in seven days, but if they are maintained at 0°C. the power of germination is retained for a longer period. The seeds show still greater vitality if they are kept at 0°C under rather dry conditions. In such circumstances germination with *S. pierotii* will occur 360 days after harvesting the seeds. Davis found that the seeds of *Cornus florida* were dormant when the fruit has first matured, and that this dormancy could not be broken by treatment with acids, ether or ethylene.

A period of after-ripening of 100 to 130 days was necessary at temperatures varying between 0°C. to 10°C. before germination would take place. During the after-ripening there was an increase in starch, sugar and amino-acid content of the seeds, but little change in fats, phosphatides or acidity. The catalase activity of the seeds was closely parallel to the after-ripening process. The seeds of Sambucus canadensis and Berberis thunbergii required alternating temperatures for germination.

Since in the seeds of Cornus florida increased capacity for germination was accompanied by increased catalase activity as well as an increase in the amount of starch, sugar and soluble protein, while the respiration fell until germination actually began, Davis suggested that the process of after-ripening was closely linked with respiration. Further evidence for this view was given by the fact that if the seeds were exposed to higher temperatures (15°C. and over) there was an increase in readily usable food material, the catalase activity fell and the respiration rose.

DEVELOPMENT OF EXPRESSION AND METABOLISM

Aided by assimilation and respiration, the seedling or sporeling is transformed into the mature individual. Growth is the final expression of successful metabolism. Without metabolism there can be no growth, and when metabolism ceases, death supervenes. It was early shown by Sachs that the only true criterion of growth is increase in dry-weight (material dried to constant weight at 100°C.), for transpiration and water absorption lead to considerable variations in fresh-weight, and no reliance can be placed on such figures. Estimation of dry-weight, however, has the disadvantage that the plant is killed, and now increase in leaf area is much used to measure growth, for it has been shown that increase in this organ is also an index of true growth.

Both the anabolic and the catabolic sides of metabolism are necessary for growth. The living plant must be supplied with food, and it must also be supplied with energy, without which its vital activities would soon cease. It is necessary, however, that the anabolic side of metabolism should be in excess of the catabolic, or no balance of material would be available for growth.

It has been shown by Boysen-Jenscn that a balance is always available under ordinary conditions for growth. He attempted to elucidate the economic working of two biologically different types of plants: Sinapis, a sun-loving plant, and Oxalis, a shade type. In Sinapis the maximum intensity of assimilation was estimated at 6 nig. of carbon dioxide per 50 cm.^{2} of leaf area every hour at 20°C. The dry-weight of a plant with a leaf area of 50 cm.^{2} was 0-279 gm., and the respiratory rate was found to be 0-8 mg. of carbon dioxide for the same leaf area and at the same temperature.

Boysen-Jensen was therefore able to calculate that the amount of dry-matter assimilated in the course of a day was 60 mg., and loss in dry-matter from respiration as 14 mg., or, in other words, the gain of material over loss was 46 mg., or 16-5 per cent Here there is an ample margin of the anabolic over the catabolic side of metabolism to allow of fresh growth. In shade plants, such as *Oocalis,* the assimilation and respiration values are considerably lower than in *Sinapis.* In *Oxalis* itself the maximal assimilation was 0-8 mg. of carbon dioxide per 50 cm^2 of leaf area every hour at 20°C., while the respiration rate varied between 0-1 to 0-2 mg. for a similar leaf area for the same time and temperature. Again, it is evident that there is a balance of material available for growth.

GROWTH FACTORS

External and internal factors have a powerful influence on the growth of the living plant. If the factors are not suitable, growth may be retarded or even completely inhibited; or again, the absence of one or more factors may act in a detrimental manner on the well-being of the plant.Among the principal factors concerned in growth are food and water. Without the right kind and quantity of food the plant cannot survive for any length of time.

5

Role of Garden Pea Seeds

The garden pea (*Pisum sativum*) is a hardy, cool-season annual. It is quite variable in size, ranging from 1 to 6 feet in height, since some varieties are dwarf, others half dwarf, and still others tall. It is cultivated in nearly all home gardens, in market and truck gardens, and is also grown on a very large scale for canning. Since it does not thrive during midsummer it is grown as a partial-season crop, *i.e.*, in spring and early summer in the North and as a winter and spring crop in the South. Like other members of its family (*Leguminosae*), through the agency of bacteria in its root nodules, the pea is able to utilize nitrogen compounds which have been formed by using the free nitrogen of the soil air. Garden peas of a late or main-crop variety, known as Telephone, were planted Apr. 10. They were sown in rows 30 inches distant and the plants were later thinned to 8 inches apart in the row.

EARLY DEVELOPMENT

The plants were excavated for the first time on May 23. They were 11 inches high and the stems 5 millimeters thick. Each plant had two or three branches and about a dozen large compound leaves some of which were not fully developed. The total leaf surface was 1.5 square feet. The plants had strong taproots about 4 millimeters in diameter at their origin. These tapered to 1 millimeter in thickness below the 9-inch level. Depths of penetration of 20 to 26 inches were found. The course of the roots was often somewhat tortuous. The bulk of the absorbing area was furnished by the numerous, strong, lateral branches nearly all of which arose from the taproot in the surface soil.

Just below the seed, which was 1 inch deep, these occurred in great abundance. One rather large plant gave rise to 21 branches about 1 millimeter in diameter from the first inch; 14 nearly 1 millimeter thick from the second inch; 11 averaging ½ millimeter in diameter from the third inch; and 16 more, about 1/3 millimeter thick, in the next 6 inches. The specimen drawn had only about two-thirds as many roots. The general course of these roots was outward and only slightly downward. Thus nearly all ended in the first 6 to 9 inches of soil. A few of the longest extended laterally 15 to 18 inches and then turned downward. Depth of penetration, however, was not marked and only a few reached the 12- to 18-inch soil level. Hence, the plant had a rather superficial root system at this stage of growth. The absence of roots in the surface 1 to 2 inches of soil is of interest since it bears a direct relation to root injury by cultivation.

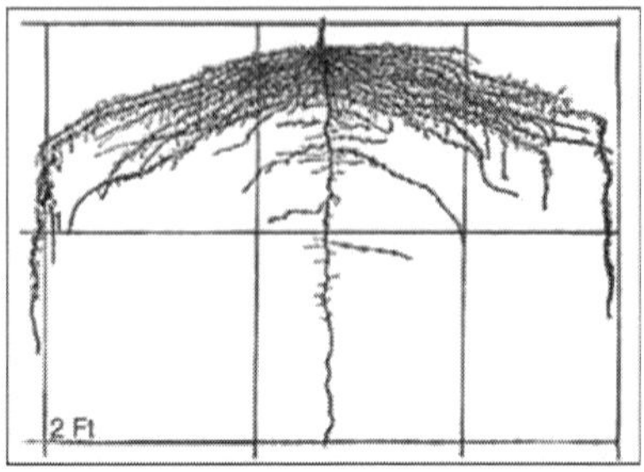

Fig. Root System o f the Telephone Garden Pea 6 Weeks Old.

As regards branching of the taproot, it may be noted that some of the laterals, even in the surface 6 inches, were only 3 to 9 inches long. Below this depth branching was much poorer as is clearly illustrated in the drawing. The length of the laterals decreased with depth of origin and the last 4 to 6 inches of the taproot were unbranched. This, of course, was also true of the major laterals. These were furnished with rootlets at the rate of two to eight per inch, distribution being somewhat variable. These sublaterals usually varied from 0.2 to 1 inch in length, although they were occasionally longer. They were rarely branched.

LATER DEVELOPMENT

A second examination was made June 17 when the peas were blossoming. There were about four branches per plant and these varied from 1 to 2.5 feet in length with an average length of 18 inches. Plants of average size possessed 45 leaves and a total leaf surface of slightly more than 5 square feet.

The pronounced taproots had increased to 7 millimeters in diameter near the soil surface. They tapered to 1.5 millimeters at a depth of 1 foot but were nearly 1 millimeter thick throughout their sinuous course. Maximum depths of 36 to 38 inches were ascertained. Branching had increased to a marked degree; a few roots 0.5 to 1 inch long now occurred in the first inch of soil. Frequently as many as 100 laterals were found to originate from a taproot in the first 12 inches of soil. A maximum of 18 per inch was determined. These were 0.5 to 2.5 millimeters in diameter. In the second and third foot-portions of the taproot which were poorly branched at the previous examination-a total of 110 to 130 rather uniformly distributed laterals frequently arose. The branches become longer, in general, on the older portions of the taproots. Near the root ends no branching occurred. The lateral spread had increased only a few inches beyond that of the previous examination (now about 22 inches) but many of the widely spreading roots had turned downward and extended well into the second foot of soil. Moreover, the branching had progressed with the growth of the main laterals. Branches occurred at a somewhat variable rate, about seven per inch of lateral. This was also the usual rate of branching for the main laterals. Usually the branches were only 0.2 to 0.5 inch long although a few of them attained a length of 2 inches. Thus new soil areas were ramified for water and nutrients.

A pronounced feature was the greater number of branches on the portions of the laterals at some distance from their origin. In fact the first few inches were often poorly branched. Branchlets of the third order were more pronounced than at the previous examination although on the newer growth only the longest secondary laterals were rebranched. No long laterals arose below the first foot of soil but the very abundant short ones (1 to 3 inches) added considerably to the absorbing area.

In summarizing, four differences were apparent from the previous root development: The taproots had increased in depth from about 2 to 3 feet and the part below 10 inches had become clothed with a large number of short branches. The number of lateral roots in the surface 12 inches of soil had greatly increased, especially on the second 6 inches of the taproot. The widely spreading and some of the obliquely penetrating laterals had turned downward and extended well into the second foot of soil. Finally, branching had greatly increased; older branches were somewhat longer and better rebranched and abundant new ones had arisen on the elongating roots. Thus the second and third 6-inch soil levels were rather thoroughly ramified.

MATURE PLANTS

A final study was made July 11. The well-branched vines, which were 40 inches tall, were beginning to dry at the base. The abundant pods bore peas which were half dry and too mature for use as green peas.The pronounced taproots were traced throughout their devious course which was usually characterized by gentle curves but often by abrupt, almost right-angled turns. Depths of 3 to 3.2 feet were found. This, however, was no greater than on June 17. The number of laterals, moreover, had not increased in the surface foot of soil.

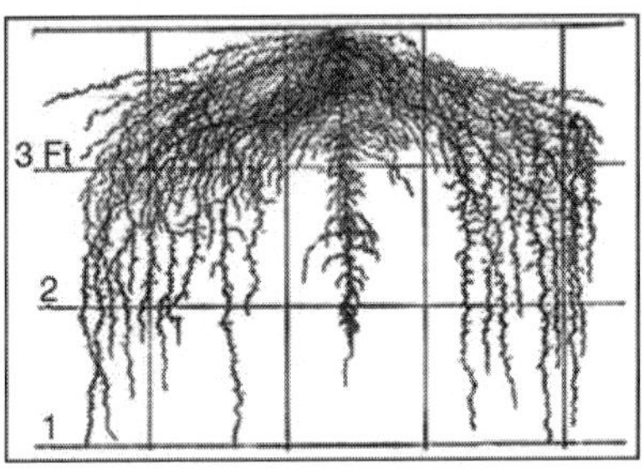

Fig. Mature Root System of the Pea. Note the Large Soil Volume Occupied and the Greater Degree of Branching than at the Earlier Examination.

The lateral spread was a little greater (maximum, 2 feet), more long secondary laterals occurred, and branches of the third order were much more frequent. In addition to the more thorough ramification of the soil already occupied, an extensive new volume of soil had been occupied by the downwardly penetrating main lateral roots. At the June examination the longest of these did not extend beyond the second foot. At this time the entire second foot was thoroughly ramified and much of the third foot was also occupied. This added greatly to the absorbing area.

It is of interest to note that some of these roots extended to greater depths (maximum, 36 inches) than the taproot. Also, their distribution was such that the soil volume below a depth of 1 foot and to 6 inches on all sides of the taproot was almost unoccupied. Whether or not this was ramified later by an elongation of the branches of the taproot was not ascertained. At this time the taproot below the first foot was furnished with relatively short branches. Only rarely did they exceed 4 inches in length and they were frequently 0.5 to 2 inches long. The number was, as before, about five per inch but they had not only grown in length but were also much more profusely rebranched. Branches of the third order occurred on some of the older and longer laterals.

This increase in length of branchlets also characterized the larger main branches. Most of the branches of the second order ranged between 0.2 and 3 inches but not infrequently sublaterals 5 to 10 or more inches long occurred. Branchlets of the third order seldom exceeded four per inch in number and ranged from 1 millimeter to 1 inch in length. In the deeper soil, especially secondary branches often took a horizontal course. The roots were rather tough, of a tan colour, and in many cases the root ends had dried.

The garden pea is characterized by a strong taproot which in its early development is profusely branched only in the first 6 inches of soil. Plants about 1.5 months old have a root depth of 2 feet. The surface soil at a depth of 2 to 8 inches is well filled with a network of nearly horizontal roots and their laterals to a distance of 18 inches on all sides of the plant. But in the deeper soil little absorbing area occurs. About a month later, when the plants are blossoming, the root system is much more extensive and efficient. The taproot is 3 feet long and much better branched throughout its entire course, secondary branches are longer and much more numerous, and branches of the third order abundant except on the youngest parts. Lateral spread has been increased to 22 inches. Many of the formerly horizontal roots have turned downward and, with those penetrating more obliquely, fairly well occupied the second foot of soil.

Nearly a month later, when the seeds are drying, the taproots have not increased in depth, nor has the number of main laterals increased in the surface soil. But the soil volume earlier delimited is much more thoroughly occupied as a result of an elongation and more profuse branching of the finer laterals. By a downward extension of the main laterals, moreover, the second foot of soil is well filled and the third foot fairly well ramified. Branches from the taproot are profuse but rather short, so that some of the deeper soil area directly beneath the plant is not fully occupied. Thus the pea completes the development of an extensive absorbing system after the beginning of blossoming.

OTHER INVESTIGATIONS ON PEAS

The taproot extended nearly perpendicularly downward to the depth of 39 inches. Below this it was too delicate to trace. Branches separated from the taproot throughout its length. These were most numerous between 4 and 8 inches in depth, where they seemed nearly to fill the soil for a distance of about 8 inches on either side. We traced a single branch root a distance of 18 inches from the taproot. The majority of the branches appeared to extend little farther than 1 foot. They gradually became shorter as the depth increased, but were 4 to 6 inches long at a depth of 30 inches. Sometimes the branches curved upward after leaving the taproot.

The American Wonder pea was also examined at the same stage of. development but the plant was only 6 inches tall. The roots extended almost exclusively downward, the taproot reaching a depth of 30 inches. No branches extended a distance greater than 4 inches from it.

Investigations in Russia indicate that although during germination and sprouting the vertical roots are rather short, later they make a uniform and rapid growth. Plants only in the second-leaf stage had a root depth of 13 inches. This was the only one of a number of vegetable crops studied that did not extend its vertical roots deeper after the inception of the flowering period. The development of the horizontal roots at the beginning was very much less than that of the vertical ones. But during the period of flowering these continued growth rapidly (after growth of the vertical roots had ceased) increasing in length from 12 to 20.5 inches. A maximum root depth of 3 feet was attained.

Numerous investigations on the pea have been made in Germany. In one it was found that it had a strong taproot that deeply penetrated with a marked early development of laterals which were especially abundant on the upper portion of the root system. These were found to extend in an obliquely outward and then downward course, many of them being nearly as long as the main root. Branching was profuse. In a good loam soil the Victoria pea was found to reach a depth of 18 inches and a spread of 12 inches when only 36 days old. Other investigations reveal a similar root habit, plants 1 foot tall having taproots 28 inches deep. The laterals attained a length of 10 inches. The upper laterals and their branches were usually much better developed than the lower ones. Occasionally, however, single branches arising deeply from the taproot made a good growth. Upon the loss of its tip, the taproot became covered with very numerous, long, much-branched laterals. These were augmented by a good adventitious root formation. Upon injury to the end of the taproot the laterals extended deeper than usual, those arising from near the injured end of the taproot growing more obliquely downward. Sometimes a single strong lateral turned vertically downward and, deeply penetrating, took the place of the taproot.

In another experiment peas of the Victoria variety were grown in filled pits in a field of well-compacted loam. The soil was washed from the roots at various stages of their development. At the age of 32 days the roots of the seedling plants were more than 10 inches deep, 56 inches at the beginning of blossoming, 69 inches at the beginning of fruiting, and about 7 feet deep when the fruits were ripe. Thus the growth in depth increased rapidly at about blossoming time and continued until maturity of the fruit. Moreover, the later root development consisted largely in an increase in root depth and not in lateral spread. Nodules occurred to a depth of 67 inches but the bulk were in the upper 6.5-inch layer of soil. Later studies confirmed the deeply rooting habit.

Another German investigator found that the pea was very similar to the common bean in its root habit. Like the bean, it formed a clearly defined taproot and then a row of adventitious roots which arose at first from the base of the hypocotyl but later, in considerable numbers, from higher parts of the base of the stem. These roots spread widely. Depths of 28 to over 31 inches were attained.89 In another study it was found that the pea was most profusely branched to a depth of 4 inches and that the largest laterals had a length of 12 inches. The number of laterals of the first order was about six per centimeter of taproot in the upper portion of the root system but only three on the deeper part. Laterals of the second order varied from two to three per centimeter. The deeper portion of the root system was only 52 per cent as well branched as the shallower part. Other European investigators have obtained similar results.

Field peas, of the variety *arvense,* were examined at Fargo, N. D., 86 days after planting and when the seed was ripe. They showed a sparse growth of roots in comparison to top development. The vines were 5.5 feet long but the roots reached a depth of only 3 feet and had rather a scanty supply of branches. The bulk of the roots was found within 8 to 10 inches of the surface.

Recent experiments with peas at Greeley, Colo., where a dwarf variety (Nott's Excelsior) with a short root system was crossed with a tall variety (Telephone) with a deep root system, indicate that root characters are hereditary and segregate out in the *F*2 generation according to the Mendelian ratios. A repetition of the experiment under somewhat different conditions of growth confirmed the results, but it was also found that the dwarf variety had almost as long a root system as the taller one. 67 This illustrates the fallacy of judging root extent by top growth. After years of study of scores of native and cultivated plants, it has been fully demonstrated that such a criterion is entirely untrustworthy.

A study of these investigations on different varieties of peas supports the conclusion that this species has a somewhat deeply rooting habit of growth. The lateral spread is also similar to that already illustrated and branching is quite profuse throughout. Considerable root growth after the time of blossoming was found by most investigators; in some varieties this growth seemed to be chiefly that of the branches, in others an extension of the taproot. Further studies, including adaptation of root system to different kinds of soil, are needed.

ROLE OF ROOT TUBERCLES

Peas and other members of the family of legumes are the most important plants, although not the only ones, that develop nodules or root tubercles. Various strains 96 of a certain motile bacterium (*Pseudomonas radicicola*) live in the soil. They gain entrance to the root system through the root hairs and push their way back into the cortical tissue of the root where they increase rapidly and occur in great numbers. As a result of their activities, the cells of the cortex of the host make an abnormal growth which results in the well-known enlargements called *tubercles.* These frequently occur at great depths. On native legumes they have been repeatedly observed at depths of 10 to 13 feet. Although usually most abundant on roots of garden crops in the surface 8 to 16 inches of soil, they are frequently found irregularly distributed over the root system at depths of several feet.

The bacteria secure their supplies of water, carbohydrates, etc. from the host plants but build up nitrogen compounds for which the source of nitrogen is the soil air. From these compounds the legume secures large amounts of valuable nitrogenous material. When the nodules decay, due to the activities of several other kinds of bacteria, the protein contents undergo various chemical changes and are finally left in the soil as nitrates, a most favourable source of nitrogen for green plants. From 40 to over 250 pounds of nitrogen per acre may thus be added to the soil in a single season through the activities of tubercle-forming bacteria.

All parts of legumes are comparatively rich in proteins and are very valuable as fertilizers. This explains why the practice of growing leguminous crops and plowing them under has such a stimulating effect upon the growth of succeeding crops. On raw soils low in nitrates, such as railway cuts and embankments, where other plants can scarcely grow, certain leguminous plants, such as sweet clover, often thrive. On the other hand, in a soil that is too rich nodule formation is not promoted.

Experiments with the development of nodules on field peas and. other legumes have shown that they become much larger when soil temperatures are most favourable. A consistent increase in dry weight of nodules occurred as the soil temperature increased from 12 to about 24°C. At higher temperatures (about 27 to 30°C. for peas) a progressive decrease occurred. 75

Likewise nodule production decreases as soil moisture diminishes below an optimum, or may entirely cease in soils that are very dry. A soil environment favourable to the growth of roots is also favourable to the growth of nodule-forming and many other species of bacteria which promote crop growth. Thus the promotion of proper soil aeration, water content, fertility, and temperature, so far as this is possible, affects plant growth not only directly by promoting root development and activities but also indirectly through its influence upon the activities of bacteria.

ROOT DEVELOPMENT IN RELATION TO CULTURAL PRACTICE

It would seem that plants with vigorously developed and quite extensive root systems like those of the pea would thrive on many kinds of soil. In fact this has been ascertained to be true. Undoubtedly the superficial and deeper portions are modified, respectively, so as best to adapt the plant to a particular environment. Such modifications have been found for beans, the plants of which are closely related and of similar general root habit. For early crops sandy loams are preferred because they are easily warmed. But they do not retain the moisture and are often less fertile than heavier-soil types such as clay and silt loam. The roots need good aeration and the soils must be well drained. If the soil is too rich, the plants will develop large vines and ripening will be delayed.

Thorough soil preparation is especially important where the crop is sown so thickly that cultivation is not possible. This is true for a part of the market crops and the canning crop which is sown broadcast or in close drills. Poor seed-bed preparation may result in markedly decreased yields. Otherwise the root habit would indicate that beneficial results will be obtained by frequent but shallow cultivation until the roots thoroughly ramify the soil and the vines cover the ground. The practice of deep planting permits the roots to start in moist, cool soil. But it may increase the prevalence of root rot, a disease caused by a soil-borne fungus (*Thieldvia basicola*), which grows best and does greater harm when more of the stem occurs below the soil. Because of the activities within the root tubercles, peas do not require such fertile soils or as heavy applications of manure as many other crops. In fact large applications of manure do not increase the yields at the same rate as smaller ones.

6

Structure of Seed Chemistry

A seed is a small embryonic plant enclosed in a covering called the seed coat, usually with some stored food. It is the product of the ripened ovule of gymnosperm and angiosperm plants which occurs after fertilization and some growth within the mother plant.

The formation of the seed completes the process of reproduction in seed plants (started with the development of flowers and pollination), with the embryo developed from the zygote and the seed coat from the integuments of the ovule.

Seeds have been an important development in the reproduction and spread of flowering plants, relative to more primitive plants like mosses, ferns and liverworts, which do not have seeds and use other means to propagate themselves. This can be seen by the success of seed plants (both gymnosperms and angiosperms) in dominating biological niches on land, from forests to grasslands both in hot and cold climates.

The term *seed* also has a general meaning that predates the above — anything that can be sown *i.e.* "seed" potatoes, "seeds" of corn or sunflower "seeds". In the case of sunflower and corn "seeds", what is sown is the seed enclosed in a shell or hull, and the potato is a tuber.

SEED STRUCTURE

A typical seed includes three basic parts:

- An embryo,
- A supply of nutrients for the embryo,
- A seed coat.

The embryo is an immature plant from which a new plant will grow under proper conditions. The embryo has one cotyledon or seed leaf in monocotyledons, two cotyledons in almost all dicotyledons and two or more in gymnosperms.

The radicle is the embryonic root. The plumule is the embryonic shoot. The embryonic stem above the point of attachment of the cotyledon(s) is the epicotyl. The embryonic stem below the point of attachment is the hypocotyl.

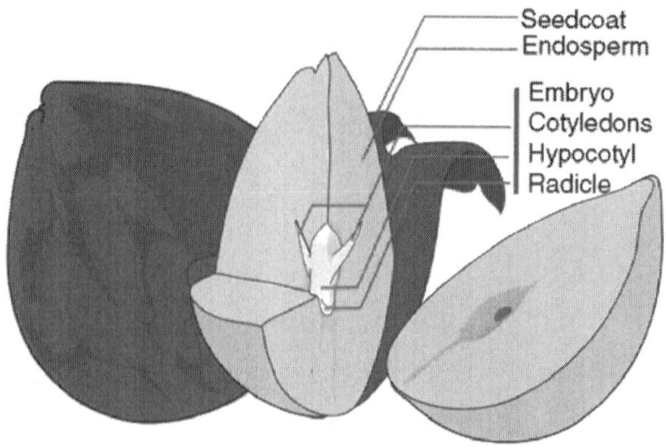

Fig. The Parts of an Avocado Seed (a dicot), Showing the seed Coat, endosperm, and embryo.

Within the seed, there usually is a store of nutrients for the seedling that will grow from the embryo. The form of the stored nutrition varies depending on the kind of plant. In angiosperms, the stored food begins as a tissue called the endosperm, which is derived from the parent plant via double fertilization. The usually triploid endosperm is rich in oil or starch and protein. In gymnosperms, such as conifers, the food storage tissue is part of the female gametophyte, a haploid tissue. In some species, the embryo is embedded in the endosperm or female gametophyte, which the seedling will use upon germination. In others, the endosperm is absorbed by the embryo as the latter grows within the developing seed, and the cotyledons of the embryo become filled with this stored food.

At maturity, seeds of these species have no endosperm and are termed exalbuminous seeds. Some exalbuminous seeds are bean, pea, oak, walnut, squash, sunflower, and radish. Seeds with an endosperm at maturity are termed albuminous seeds. Most monocots (*e.g.* grasses and palms) and many dicots (*e.g.* brazil nut and castor bean) have albuminous seeds. All gymnosperm seeds are albuminous. The seed coat (or *testa*) develops from the tissue, the integument, originally surrounding the ovule. The seed coat in the mature seed can be a paper-thin layer (*e.g.* peanut) or something more substantial (*e.g.* thick and hard in honey locust and coconut). The seed coat helps protect the embryo from mechanical injury and from drying out. In addition to the three basic seed parts, some seeds have an appendage on the seed coat such an aril (as in yew and nutmeg) or an elaiosome (as in Corydalis) or hairs (as in cotton). There may also be a scar on the seed coat, called the *hilum*; it is where the seed was attached to the ovary wall by the funiculus.

SEED DEVELOPMENT

Following the fertilizations in the embryo sac, the zygote divides repeatedly by mitosis and differentiates into an embryo. The endosperm nucleus also divides by mitosis and forms the endosperm tissue, which provides food for the developing embryo. The early embryo is linear with apical meristems on either end and one or two seed leaves or cotyledons. The axis below the cotyledons is called the hypocotyl, at the tip of which is the radicle that gives rise to the primary root of the seedling. The axis above the attachment of the cotyledons is the epicotyl, which also ends in an apical meristem. In some seeds, the first foliage leaves are formed in the seed.

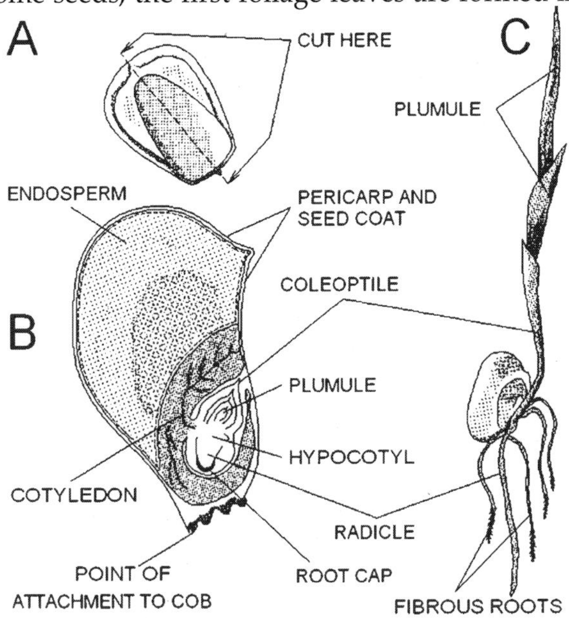

Fig. Seed Structure Monocot.

The area above the cotyledons is thus a miniature shoot and is called the plumule. In some taxa, food for the embryo remains within endosperm tissue and the cotyledons serve as organs of absorption. In others, food moves directly into the embryo and is stored within the cotyledons, leaving only a minuscule endosperm. In still another variation, the nucellus (the megasporangium wall) enlarges and becomes a storage tissue called perisperm. The integuments harden into the seed coat as the embryo matures. The scar left on the seed coat by the separation of the funiculus from the integuments is called the hilum. Often the micropyle remains visible near the hilum.

Angiosperms traditionally have been separated into two major categories on the basis of the number of cotyledons they possess monocots (mono = one; cotyledons = seed leaves) and dicots (di = two; cotyledons = seed leaves). Monocots are the grasses, sedges, lilies, and their relatives, while dicots constitute the rest of the flowering plants.

This artificial separation currently is being replaced by a more natural classification in which the *monocots* are retained as a natural group, but the dicots are separated into the eudicots (eu = true, dicots) and the magnoliids. The latter is a small group of very primitive angiosperms ancestral to both monocots and eudicots and has both woody representatives— woody magnoliids (trees like *Magnolias*, tulip trees, and laurels)—and paleoherbs (herbaceous plants such as members of the pepper and water lily families). The eudicots constitute about 97 per cent of the angiosperms, while the magnoliids make up the other 3 per cent.

In monocot seeds the single cotyledon usually digests and absorbs food from the endosperm and translocates it to the embryo. In grasses, the cotyledon is called the scutellum. In grasses, also, there is a protective sheath called the coleoptile over the plumule and another, the coleorhiza, surrounding the radicle.

Seed Germination

Internal signals shut off growth of the embryo at a certain size and the seed goes into a period of dormancy. In some plants, dormancy lasts only as long as it takes the seed to be dispersed from the ovary. In others, dormancy may last for long periods until either external or internal signals (or a combination of both) initiate further growth. Environmental factors of chief importance to initiate growth are water, light, and temperature.

Seedling Growth

When an embryo resumes growth, stored food provides the energy for seedling development—the roots first, followed by elongation of the photosynthetic shoots.

Eudicot Development

Aboveground growth in eudicots takes one of two general patterns epigeous or hypogeous. In epigeous growth, the hypocotyl elongates, pulling the plumule and cotyledons above ground; in hypogeous growth, the cotyledons remain below ground because the epicotyl grows faster than the hypocotyl and pulls the plumule erect.

Monocot Development

Monocots develop with two different general patterns one for the grasses, one for the rest of the group. In most of the monocots (but not grasses), after the radicle has pushed out of the seed coat, the first shoot structure to emerge is the cotyledon, which arches upward with the remainder of the endosperm and the seed coat still attached. It elongates above ground and is photosynthetic until the true leaves develop.

In the grasses, the sheaths around both the shoot and the root tip must be penetrated by the roots and the shoots. The root sheath, the coleorhiza, grows faster than the radicle for a short time, but when it stops growing, the radicle emerges and forms an anchoring primary root. The shoot sheath, the coleoptile, moves upward to the soil surface through elongation of the first internode of the stem (called the *mesocotyl*), and when it reaches the surface it stops growing. The plumule then pushes through into the air. At about the same time, buds of adventitious roots begin to grow and, by the time the seedling is erect with a few true leaves, it already has adventitious roots growing downward from its first node. The primary root system is short lived and dies soon after its establishment and the adventitious root system becomes the principal absorbing and anchoring system for the new grass plant.

Germination

Germination is the process whereby growth emerges from a period of dormancy. The most common example of germination is the sprouting of a seedling from a seed of an angiosperm or gymnosperm.

Fig. Sunflower Seedlings, Just three days after Germination

However, the growth of a sporeling from a spore, the growth of hyphae from fungal spores, is also germination. In a more general sense, germination can imply anything expanding into greater being from a small existence or germ.

Pollen Germination

Another germination event during the life cycle of gymnosperms and flowering plants is the germination of a pollen grain after pollination. Like seeds, pollen grains are severely dehydrated before being released to facilitate their dispersal from one plant to another. They consist of a protective coat containing several cells (up to 8 in gymnosperms, 2-3 in flowering plants). One of these cells is a tube cell.

Once the pollen grain lands on the stigma of a receptive flower (or a female cone in gymnosperms), it takes up water and germinates. Pollen germination is facilitated by hydration on the stigma, as well as the structure and physiology of the stigma and style. Pollen can also be induced to germinate *in vitro* (in a petri dish or test tube).

During germination, the tube cell elongates into a pollen tube. In the flower, the pollen tube then grows towards the ovule where it discharges the sperm produced in the pollen grain for fertilization. The germinated pollen grain with its two sperm cells is the mature male microgametophyte of these plants.

Self-incompatibility

Since most plants carry both male and female reproductive organs in their flowers, there is a high risk for self-pollination and thus inbreeding. Some plants use the control of pollen germination as a way to prevent this selfing. Germination and growth of the pollen tube involve molecular signaling between stigma and pollen. In self-incompatibility in plants, the stigma of certain plants can molecularly recognize pollen from the same plant and prevents it from germinating.

Spore Germination

Germination can also refer to the emergence of cells from resting spores and the growth of sporeling hyphae or thalli from spores in fungi, algae, and some plants.

Resting Spores

In resting spores, germination involves cracking the thick cell wall of the dormant spore. In zygomycetes the thick-walled zygosporangium cracks open and the zygospore inside gives rise to the emerging sporangiophore. In slime molds, germination refers to the emergence of amoeboid cells from the hardened spore. After cracking the spore coat, further development involves cell division, but not necessarily the development of a multicellular organism.

Zoospores

In motile zoospores, germination frequently means a lack of motility and changes in cell shape, which allow the organism to become sessile.

Ferns and Mosses

In plants such as bryophytes, ferns, and a few others, spores germinate into independent gametophytes. In the bryophytes (*e.g.* mosses and liverworts), spores germinate into protonemata, similar to fungal hyphae, from which the gametophyte grows. In ferns, the gametophytes are small, heart-shaped prothalli that can often be found underneath a spore-shedding adult plant.Seedling

A seedling is a young plant sporophyte developing out of a plant embryo from a seed. Seedling development starts with germination of the seed. A typical young seedling consists of three main parts the radicle (embryonic root), the hypocotyl (embryonic shoot), and the cotyledons (seed leaves). The two classes of flowering plants are distinguished by their numbers of seed leaves Monocotyledons (monocots) have one blade-shaped cotyledon, whereas dicotyledons (dicots) have two round cotyledons. Gymnosperms are more varied. Pine seedlings have up to eight cotyledons. The seedlings of some flowering plants have no cotyledons at all. These are said to be acotyledons.

Germination and Early Seedling Development

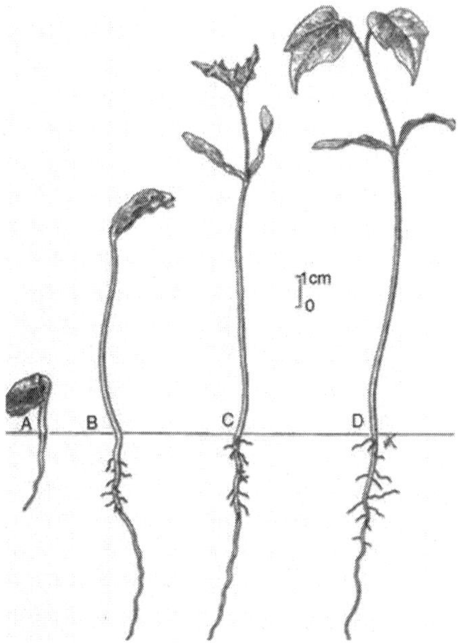

Fig. Development of an Angiosperm (maple) Seedling

During germination, the young plant emerges from its protective seed coat with its radicle first, followed by the cotyledons. The radicle orients towards gravity, while the hypocotyl orients away from gravity and elongates through cell expansion to push the cotyledons out of the ground.

PHOTOMORPHOGENESIS AND SKOTOMORPHOGENESIS

Dicot seedlings grown in the light develop short hypocotyls and open cotyledons exposing the epicotyl. This is also referred to as photomorphogenesis. In contrast, seedlings grown in the dark develop long hypocotyls and their cotyledons remain closed around the epicotyl in an *apical hook*. This is referred to as skotomorphogenesis or etiolation. Etiolated seedlings are yellowish in colour as chlorophyll synthesis and chloroplast development depend on light. They will open their cotyledons and turn green when treated with light.

In a natural situation, seedling development starts with skotomorphogenesis while the seedling is growing through the soil and attempting to reach the light as fast as possible. During this phase, the cotyledons are tightly closed and form the *apical hook* to protect the shoot apical meristem from damage while pushing through the soil. In many plants, the seed coat still covers the cotyledons for extra protection. Upon breaking the surface and reaching the light, the seedling's developmental programme is switched to photomorphogenesis. The cotyledons open upon contact with light (splitting the seed coat open, if still present) and become green, forming the first photosynthetic organs of the young plant. Until this stage, the seedling lives off the energy reserves stored in the seed. The opening of the cotyledons exposes the shoot apical meristem and the *plumule* consisting of the first *true leaves* of the young plant.

The seedlings sense light through the light receptors phytochrome (red and far-red light) and cryptochrome (blue light). Mutations in these photo receptors and their signal transduction components lead to seedling development that is at odds with light conditions, seedlings that show photomorphogenesis when grown in the dark.

Seedling Growth and Maturation

Once the seedling starts to photosynthesize, it is no longer dependent on the seed's energy reserves. The apical meristems start growing and give rise to the root and shoot. The first "true" leaves expand and can often be distinguished from the round cotyledons through their species-dependent distinct shapes. While the plant is growing and developing additional leaves, the cotyledons eventually senesce and fall off.

Consumption of Seedlings

Seedlings are commonly eaten as a health food. These seedlings are usually labeled sprouts, but in a botanical sense are actually seedlings. There is controversy in whether seedlings, or sprouts, are really worth eating. These seedlings are used many times in organic foods and are eaten for their concentrations of certain vitamins the seedling naturally contains. This can be viewed out of proportion as the seedling is usually too small to contain enough vitamins or minerals to be in line with the amount they are claimed to obtain.

Sprouting

Sprouting is the practice of soaking, draining and then rinsing seeds at regular intervals until they germinate, or sprout. This can be a semi - automated or fully automated process when done on a large scale for commercial use.

Seeds that can be Sprouted

One of the most common sprouts is that of the mung bean (*Vigna radiata*), often sold as 'Chinese Bean Sprouts'; another common sprout is the alfalfa sprout and the barley sprout. Other seeds that can be sprouted include adzuki bean, almond, amaranth, annatto seed, anise seed, arugula, basil, navy bean, pinto bean, lima bean, broccoli, buckwheat, cabbage, canola seed, caragana, cauliflower, celery, chia seed, chickpeas, chives, cilantro (coriander, dhania), clover, cress, dill, fennel, fenugreek, flax seed, garlic, hemp seed, kale, kamut, kat, leek, green lentils, lupins,pearl millet, mizuna, mustard, oats, onion, black-eyed peas, green peas, pigeon peas, snow peas, peanut, psyllium, pumpkin, quinoa, radish, rye, sesame, soybean, spelt, sunflower, tatsoi, triticale, watercress, and wheat berries.

However, many sprouts are in fact toxic when eaten, like kidney beans. Some sprouts can be cooked to remove the toxin, while others will be toxic either way and should be avoided. So before eating any sprouts, find out if that species is edible as a sprout. With all seeds, care should be taken that they are intended for sprouting or human consumption rather than sowing. Seeds intended for sowing may be treated with chemical dressings. Several countries, such as New Zealand, also require that some varieties of edible seed be heat-treated, thus making them impossible to sprout.

Many varieties of nuts, such as almonds and peanuts, can also be started in their growth cycle by soaking and sprouting, although because the sprouts are generally still very tiny when eaten, they are usually called "soaks."

Moisture, warmth, and in most cases, indirect sunlight are necessary for sprouting. Some sprouts, such as mung beans, can be grown in the dark. Little time, effort or space is needed to make sprouts.

To sprout seeds, the seeds are moistened, then left at room temperature (between 13 and 21 degrees Celsius) in a sprouting vessel. Many different types of vessels can be used. One type is a simple glass jar with a piece of cloth secured over its rim. 'Tiered' clear plastic sprouters are commercially available, allowing a number of "crops" to be grown simultaneously. By staggering sowings, a constant supply of young sprouts can be ensured. Any vessel used for sprouting must allow water to drain from it, because sprouts that sit in water will rot quickly. The seeds will swell and begin germinating within a day or two.

Sprouts are rinsed as little as twice a day, but possibly three or four times a day in hotter climates, to prevent them from souring. Each seed has its own ideal sprouting time. Depending on which seed is used, after three to five days they will have grown to two or three inches in length and will be suitable for consumption.

If left longer they will begin to develop leaves, and are then known as baby greens. A popular baby green is sunflower after 7-10 days. The growth process of any sprout can be slowed or halted by refrigerating until needed.

Common causes for sprouts to turn out inedible

- Seeds are allowed to dry out
- Seeds are left in standing water
- Temperature is high or too low
- Insufficient rinsing

- Dirty equipment
- Insufficient air flow
- Contaminated source of water
- Poor rate of germination of seed

These problems are easily solved by an automatic sprouter that mists and drains the sprouts at regular intervals. To control temperature, in the winter a warming blanket can be placed under the sprouter, and in the summer small fans in the lid if it's very hot and humid.

Mung beans can be sprouted either in light or dark conditions. Those sprouted in the dark will be crisper in texture and whiter, as in the case of commercially available Chinese Bean Sprouts, but these have less nutritional content than those grown in partial sunlight.

Growing in full sunlight is not recommended, because it can cause the beans to overheat or dry out. Subjecting the sprouts to pressure, by placing a weight on top of them in their sprouting container, will result in larger, crunchier sprouts similar to those sold in Polish grocery stores.

A very effective way to sprout beans like lentils or adzuki is in colanders. Soak the beans in hot water for about 8 hours then place in the colander. Wash twice a day. The sprouted beans can be eaten raw or cooked.

NUTRITIONAL INFORMATION

Sprouts are rich in digestible energy, bioavailable vitamins, minerals, amino acids, proteins, beneficial enzymes and phytochemicals, as these are necessary for a germinating plant to grow. These nutrients are essential for human health. To clarify the nutritional changes upon germination & sprouting have been summarized below.

Chavan and Kadam concluded that - "The desirable nutritional changes that occur during sprouting are mainly due to the breakdown of complex compounds into a more simple form, transformation into essential constituents, and breakdown of nutritionally undesirable constituents."

"The metabolic activity of resting seeds increases as soon as they are hydrated during soaking. Complex biochemical changes occur during hydration and subsequent sprouting. The reserve chemical constituents, such as protein, starch and lipids, are broken down by enzymes into simple compounds that are used to make new compounds."

"Sprouting grains causes increased activities of hydrolytic enzymes, improvements in the contents of total proteins, fat, certain essential amino acids, total sugars, B-group vitamins, and a decrease in dry matter, starch and anti-nutrients. The increased contents of protein, fat, fibre and total ash are only apparent and attributable to the disappearance of starch.

However, improvements in amino acid composition, B-group vitamins, sugars, protein and starch digestibilities, and decrease in phytates and protease inhibitors are the metabolic effects of the sprouting process."

INCREASES IN PLANT ENZYME CONTENT

Sprouts are a tremendous source of (plant) digestive enzymes. Enzymes act as biological catalysts needed for the complete digestion of protein, carbohydrates & fats. The physiology of vitamins, minerals and trace elements is also dependant on enzyme activity. "Being eaten whilst extremely young, "alive" and rapidly developing, sprouts have been acclaimed as the "most enzyme-rich food on the planet". Estimates suggest there can be up to 100 times more enzymes in sprouts than in fruit and vegetables, depending on the particular type of enzyme and the variety of seed being sprouted. The period of greatest enzyme activity in sprouts is generally between germination and 7 days of age."

"Grains and legume seeds of all plants contain abundant enzymes. However, while grains and seeds are dry, enzymes are largely inactive, due to enzyme inhibitors, until given moisture to activate germination. It is these inhibitors that enable many seeds to last for years in soil without deteriorating, whilst waiting for moisture. Enzyme inhibitors in some grains and legume seeds need to be inactivated by heating or other processes, before they can be safely fed. However, heating, cooking and grinding processes can also inactivate certain digestive enzymes within grains and seeds. Fortunately, during germination and sprouting of grains and seeds, many enzyme inhibitors are effectively neutralized, whilst at the same time the activity of beneficial plant digestive enzymes is greatly enhanced."

Increases in Crude Protein content Morgan et al. found that - "The protein content of sprouts increased from the time of germination. The absorption of nitrates facilitates the metabolism of nitrogenous compounds from carbohydrate reserves, thus increasing crude protein levels." Increases in Protein Quality Chavan and Kadam stated - "Very complex qualitative changes are reported to occur during soaking and sprouting of seeds. The conversion of storage proteins of cereal grains into albumins and globulins during sprouting may improve the quality of cereal proteins. Many studies have shown an increase in the content of the amino acid Lysine with sprouting." "An increase in proteolytic activity during sprouting is desirable for nutritional improvement of cereals because it leads to hydrolysis of prolamins and the liberated amino acids such as glutamic and proline are converted to limiting amino acids such as lysine." Increases in Crude Fibre content Cuddeford, based on data obtained by Peer and Leeson, stated - "In sprouted barley, crude fibre, a major constituent of cell walls, increases both in percentage and real terms, with the synthesis of structural carbohydrates, such as cellulose and hemicellulose".

INCREASES IN ESSENTIAL FATTY ACIDS

An increase in lipase activity has been reported in barley by MacLeod and White, as cited by Chavan and Kadam. Increased lipolytic activity during germination and sprouting causes hydrolysis of triacylglycerols to glycerol and constituent fatty acids. Sprouting treatment of cereal grains generally improves their vitamin value, especially the B-group vitamins. Certain vitamins such as á-tocopherol (Vitamin-E) and â-carotene (Vitamin-A precursor) are produced during the growth process.

Sprouts provide a good supply of Vitamins A, E and C plus B complex. Like enzymes, vitamins serve as bioactive catalysts to assist in the digestion and metabolism of feeds and the release of energy. They are also essential for the healing and repair of cells. However, vitamins are very perishable, and in general, the fresher the feeds eaten, the higher the vitamin content.

The vitamin content of some seeds can increase by up to 20 times their original value within several days of sprouting. Mung Bean sprouts have B vitamin increases, compared to the dry seeds, of - B1 up 285 per cent, B2 up 515 per cent, B3 up 256 per cent. Even soaking seeds overnight in water yields greatly increased amounts of B vitamins, as well as Vitamin C. Compared with mature plants, sprouts can yield vitamin contents 30 times higher."

Chelation of Minerals Shipard claims that - "When seeds are sprouted, minerals chelate or merge with protein, in a way that increases their function."

Risks and Antinutritional factors Some legumes can contain toxins or antinutritional factors, which can be reduced by soaking, sprouting and cooking (eg, stir frying). Joy Larkcom advises that to be on the safe side "one shouldn't eat large quantities of raw legume sprouts on a regular basis, no more than about 550g daily". However not all legume sprouts contain these antinutritional factors and many have greatly beneficial properties so it is recommended that the advice of a qualified nutritionist is sought before making any decisions about what to include or eliminate from a diet/ration.

Reduction of Anti-Nutritional Factors Phytic Acid occurs primarily in the seed coats and germ of plant seeds. It forms insoluble or nearly insoluble compounds with minerals including Calcium, Iron, Magnesium and Zinc, such that they cannot be effectively absorbed into the blood. Diets high in phytic acid and poor in these minerals produce mineral deficiency symptoms in experimental animals.

The latter authors state that the sprouting of cereals has been reported to decrease levels of Phytic Acid. Similarly, Shipard states that enzymes of germination and sprouting have the ability to eliminate detrimental substances such as Phytic Acid.

Buckwheat greens contain fagopyrin, a naturally occurring substance in the buckwheat plant. When ingested in sufficient quantity, fagopyrin is known to cause the skin of animals and people to become phototoxic, which is to say hypersensitive to sunlight, particularly if juiced or eaten in large quantities. Due to the growing popularity of sprouts in general, and a widespread ignorance as to the toxic dangers posed by buckwheat greens specifically, many people are today suffering unnecessarily.

SPROUTING AND THE LIVING FOODS DIET

Advocates of a raw food diet promote the use of sprouting as an effective way to increase the nutrient value, and digestibility, of beans, seeds and nuts. Sprouts are in fact the most nutrient dense food on earth–nutrient density refers to the amount of nutrients per calorie. This is why many people on a low-calorie diet make sprouts and baby greens the mainstay of their diet, especially blended for easier digestion.

Many raw food dietitians avoid grains in favour of nuts and seeds. While raw nuts and seeds contain enzyme deactivators that harm the stomach and cooking nuts and seeds denatures their fats and oils, soaked or sprouted nuts and seeds have natural oils and active enzymes at the same time.

One main criticism of the raw foods diet, especially the insistence on the health value of sprouted nuts, seeds, and occasionally grains, is the very intensive labour time require for food preparation. Not only does it take a long time and lots of labour to sprout nuts and seeds, but the sprouting process makes them not as shelf-stable.

Furthermore, unlike most cooked foods, there are very few commercial avenues for purchasing such foods. Most "raw foods bars" are raw but not sprouted (and therefore do not have active enzymes) because the bars would not keep as long on the shelf. The same is true of most nut and seed butters.

A seed is essentially a baby in a suitcase carrying its lunch. "Baby" refers to the embryo, or immature plant, that will grow and develop into the seedling and ultimately the mature plant. The "suitcase" is the seed coat (or testa) that surrounds the seeds and "lunch" refers to the nutritive source for the germinating seedling.

The food for the germinating seedling may be stored in part of the embryo itself, such as the fleshy cotyledons of a bean seed, or it may take other forms including endosperm, which is a special starch-rich storage tissue that surrounds the embryo.

A seed is officially considered to have germinated when the young root, called the radicle, emerges from the seed coat. To germinate, a seed requires three things – water, oxygen, and a suitable temperature. Water uptake, also called imbibition, is the first stage of seed germination. During this process the dry seed, which typically has a water content of less than 10 per cent, absorbs water and swells.

This process serves to hydrate the dry components of the seed and active the metabolic machinery necessary for germination. Among the early metabolic activities occurring in the seed is the breakdown of starches stored in the seed into simple sugars that can be used for energy and building blocks for necessary cellular structures. Except for the first half-hour or so of germination when little oxygen is present, seed germination and subsequent seedling growth requires oxygen. It is required, in large part, for use in the cellular structures, called mitochondria, to produce ATP (energy).

A suitable temperature is necessary to optimize the metabolic reactions required for germination. The seeds of every species have an optimal temperature for germination; some species, such as the gourds and squashes prefer warm temperatures while other species such as radish can tolerate cooler temperatures for germination.

A seed that has not germinated because it is lacking one or more of the necessary requirements for germination is termed quiescent. These seeds are simply "resting", waiting for the appropriate conditions for germination. Given water, oxygen and/or a suitable temperature, a quiescent seed will germinate. However, even if given the proper conditions, a seed may not germinate. These seeds may fail to germinate because the seed is either dormant or "dead".

Dormant seeds have the potential to germinate but are prevented from doing so by some mechanism. Thus, even though all the proper growth conditions are present, they don't germinate unless they have been "primed" and there dormancy mechanism has been overcome. There are many dormancy mechanisms in seeds.

When some seeds, like hemp, are shed they have immature embryos that will not germinate until they undergo a period of development (called after-ripening). Other seeds, like apple, require a cold treatment, called stratification, for germination.

Many of our native plant seeds must be stratified. Some seeds have a hard seed coat that needs to be nicked (called scarification) for germination. This usually occurs as the result of natural freeze-thaw cycles.

Still other seeds require a period of heat in order to germinate. Many of these species are winter annuals that germinate in the late summer/early fall. Ultimately, the function of these varied dormancy mechanisms is to enable the seed time to disperse from the parent plant and to avoid germinating during unfavourable weather.

Humans have attempted to breed dormancy mechanisms from our crop plants. Although an advantage for a wild plant, dormancy is a problem if a farmer who want the crop to germinate uniformly and immediately upon planting. It's not easy to tell if a seed is "dead". Only if it fails to germinate when provided the proper conditions and any dormancy mechanisms are broken can we consider a seed "dead".

Seed Germination

Plants produce many fruits and therefore many seeds. This is because only a few of these seeds will find suitable places and conditions in which to germinate. The majority of the seeds will be wasted; some wind-dispersed seeds could fall in water or on stones.

When fruits and seeds are dispersed the seeds have very little water in them. The outer covering, the testa, is hard and sometimes rough and creased.

Although the seeds seem dry and inactive they are very much alive. They are said to be in a dormant state and some seeds can remain like this for a long period of time. Inside the testa there is just enough water to keep the embryo (the baby plant) living until the external conditions are correct for the seed to germinate.

The Structure of a Haricot Bean Seed

The outer covering of a seed is called the testa. It is usually hard and protects the softer parts of the seed. There is a tiny hole in the testa called the micropyle. When the seed is ready to germinate, water is taken in through the micropyle. The first root, called the radicle, will grow out of the seed through the micropyle. On the surface of the testa there is a scar which shows the place where the seed was attached to the fruit.

A dried seed is difficult to study. If the seed is soaked in water for twelve hours it takes in some of the water and swells. Peas and beans are useful seeds to study after they have been soaked because they are big. The testa can be removed easily and the inside of the seed can be examined under the binocular microscope.

Inside the Seed

Inside both the pea and the bean there are two large food-storing structures called cotyledons. The food contained in the cotyledons gives energy to the embryo when the seed germinates. The embryo is found between the two cotyledons. It is made up of two parts; the radicle, or first root, which points toward the micropyle, and the plumule, or first shoot. Under the binocular microscope the plumule can be seen as two very tiny leaves. Some types of seeds have only one small cotyledon.

Most cereal grains, such as wheat, are like this. The embryo is surrounded by a structure called the endosperm which is also a food storage area. Both the cotyledon and the endosperm will provide energy to the embryo during germination.

Seed Germination

A seed certainly looks dead. It does not seem to move, to grow, nor do anything. In fact, even with biochemical tests for the metabolic processes we associate with life (respiration, etc.) the rate of these processes is so slow that it would be difficult to determine whether there really was anything alive in a seed. Indeed if a seed is not allowed to germinate (sprout) within some certain length of time, the embryo inside will die.

Each species of seed has a certain length of viability. Some maple species have seeds that need to sprout within two weeks of being dispersed, or they die. Some seeds of Lotus plants are known to be up to 2000 years old and still can be germinated.

Assuming the seed is still viable, the embryo inside the seed coat needs something to get its metabolism actived to start the embryo growing. The process of getting a seed to germinate can be simple or complicated, and this our present subject.

Seeds Lacking True Dormancy

Common vegetable garden seeds generally lack any kind of dormancy. The seeds are ready to sprout. All they need is some moisture to get their biochemistry activated, and temperature warm enough to allow the chemistry of life to proceed. Seeds taken from the wild, however, are frequently endowed with deeper forms of dormancy.

Thick Seed Coat

Many kinds of seeds have very thick seed coats. These obviously keep water out of the seed, so the embryo cannot get the water needed to activate its metabolism and start growing. The lotus seeds are an example of this. An outstanding example from the northern temperate zone is the Kentucky coffee tree (*Gymnocladus dioica*). The seed coat is perhaps two millimeters thick! The Kentucky coffee tree holds its seed pods in the the top of the tree all winter. The inside of the pod is fleshy (lots of water). The pods are very dark in colour. Other species might use some pounding along a river or drop seeds into seacoast surf to abrade the thick seed coat. Some of the sea beans do this. Other seeds might need an vertebrate or other animal to attack the seed coat (but give up trying to eat the seed) and thereby weaken the coat. The process of nicking the thick seed coat to initiate germination is called scarification.

A final, and very common, example of a way to scarify a seed coat is observed in strawberry and raspberry. The thick seed coat is designed to be swallowed by the frugivore. The animal digests the fruit pulp, but the seed coat passes through the digestive system still protecting the viable embryo inside, but weakened enough to allow sprouting! The seed is deposited with a little organic fertilizer in the environment and can now sprout!

Thin Seed Coat

A thin seed coat is so thin that it is no barrier to water. Some other kind of dormancy mechanism is needed. Knowing that light can penetrate thin layers of plant tissue should give us the idea that light might be a signal. That plants can absorb light and respond biochemically is a fact we know from our study of photosynthesis. All we need is a pigment molecule that can absorb light and cause a change in the behaviour of the embryo.

The pigment is phytochrome. Like chlorophyll, it is made of a chromophore with tetrapyrole structure and is associated with proteins. This pigment is different from chlorophyll, however, in one critical way. It exists in two inter-conver-tible forms.

One form of phytochrome, named Pfr, is the form of the phytochome found in plant cells that are exposed to red (660 nm) or common white light. This form of phytochrome is biologically very active and plays a role in all systems when a plant needs to know if the lights are "on" or "off." In lettuce (*Lactuca sativa*) seeds, Pfr causes the seeds to begin to germinate as we will soon see. Thus lettuce seeds germinate only when placed in white or red light. Buried in deep soil, they will not germinate.

The other form of phytochrome, named Pr, is formed when phytochrome is exposed to far-red (730 nm) light. This form is biologically inactive or inhibits responses. Thus if lettuce seeds are placed in far-red light they do not germinate.

Large seeds have lots of storage material. If their seed coat is very thin, their evolution may have arrived at a completely different response. Think about pea seeds. They are large and have very thin seed coats. How would they respond to light?

Insufficient Development

If a seed's embryo is not completely developed, some additional maturation may be needed before the seed can sprout. This happens in seeds with little-to-no storage material invested in the seed. Examples include orchid seeds. They are the size of dust and have almost nothing but a very immature embryo on-board.

Such a seed needs an association with fungi in the soil or other environments to feed the developing embryo until the embryo is mature enough to actually penetrate the seed coat. These seeds are also likely to have a very brief viability. The fungal association must be established rapidly or the embryo dies.

Inhibitors Present

Many plant species invest chemicals in the developing seeds, and these chemicals inhibit the development of the embryos. They keep the embryos dormant. Obviously the seed must have some way to eliminate these chemicals before they can sprout.

Abscisic Acid

Many temperate zone species that use inhibitors use abscisic acid. This chemical induces dormancy in the embryo. The chemical is produced in abundance in the late summer and early fall. The seeds in the fruits become dormant so, even if they are dispersed in autumn, they cannot sprout. During the winter enzymes in the seeds degrade the abscisic acid. By spring the abscisic acid is gone and the seed can sprout.

We can collect seeds of these species and get them to sprout early. The seeds are put in moist soil and refrigerated for about four weeks (a process often called stratification). This is sufficient time to degrade the abscisic acid. Then the planted seeds are placed in a warm greenhouse. The seeds assume winter is over, spring has come, and they begin to sprout. This process is called vernalization. If we think of "vernal" as meaning "spring" then we understand how we got this name!

PHENOLIC COMPOUNDS

Plants that live in deserts have a different problem. There is no cold, moist, winter to allow vernalization of abscisic acid. These plants instead use more potent toxins, phenolic compounds, to keep their seeds dormant until the proper season for germination. Phenolic compounds are freely water-soluble, the plant is living in a desert.

Deserts typically have very long dry seasons and a short wet season accompanied by flash floods and so on. How do we think the phenolic compounds are lost? How would the mechanism ensure that seeds do not sprout in the dry season, but only after the seed could be sure it is in the wet season? The word leaching might give we a hint?

As any "typical" seed it has three fundamental parts:

- A seed coat
- A storage area (in this case the endosperm) and
- A dormant embryo

The seed coat is really a fruit coat. In all grains, the seed coat (former ovule integument) is fused to the ovary wall (the true fruit wall). So in fact, the grain is technically a fruit (caryopsis) even though we often call it a seed.

Barley is used primarily in beer making. Brewers use barley as a source of sugar to make the alcohol for beer. Brewers knew that sprouting the barley seeds improves the sugar yields tremendously, but they wanted even more sugar yield. The brewers came to plant physiologists to find out if anything could be done.

Study of the seed showed that the seed/fruit coat is water-resistant and thereby reduces the rate of water uptake by the seed, and water uptake is essential for seed germination and the improved sugar yields.

The sugar content of a dry barley seed is actually quite low, but the endosperm holds a huge reserve of starch. Starch is a polymer of sugars. In a sense it is a long chain of sugar molecules linked together. This is probably the source of the sugar.

The barley embryo has three parts:

- The cotyledon (or seed leaf). Since there is only one, barley is a monocot.
- The epicotyl (becomes the shoot)
- The radicle (becomes the root)

Seed germination is said to have occurred when growth of the radicle bursts the seed coat and protrudes as a young root. The energy for seed germination probably comes from respiration of the sugar in the endosperm. However, the embryo and the starch are separated from each other. There must be some chemical communications that physiologists could manipulate to enhance sugar yield for the brewers.

The endosperm of the seed has two parts. The bulk of the volume is a starch storage area. The covering layer is called the aleurone layer. The aleurone is made of cells that store protein in abundance.

We probably know about this in the case of rice. Rice grains are treated before we buy them for cooking. The fruit/seed coat is milled off and discarded as it has little food value. The embryo breaks out and is called the "germ." In wheat, the embryos are sold as wheat germ.

The milling process leaves the rice endosperm surrounded by aleurone. This is sold as brown rice and has food value in starch and protein. It has an interesting texture after cooking. If the endosperm is polished before it is sold, the aleurone can be removed. Polishing results in white rice which has food value in starch only. Thus white rice, while considered much "finer" than brown rice, is actually a lower-quality food.

The first step in barley seed germination is imbibition. In this process, water penetrates the seed coat and begins to soften the hard, dry tissues inside. The water uptake causes the grain to swell up. The seed/fruit coat usually splits open allowing water to enter even faster. The water begins to activate the biochemistry of the dormant embryo.

The water coming into the seed and embryo dissolves a chemical made inside the embryo. This chemical is called Gibberellic Acid (GA). It is a plant hormone, not too different from steroids. The dissolved GA is transported with the water through the rest of the seed tissues until it arrives at the aleurone layer. The GA crosses into the cytoplasm of the aleurone cells and turns on certain genes in the nuclear DNA. DNA is, of course, the hereditary molecule and contains the instructions for making every protein needed for the survival of a barley plant.

The precise mechanism of how GA turns on the DNA is unknown at present. It is clear however that the mode of action is to turn on just *certain* genes in the DNA. The genes that are turned on are transcribed. The information archived in the DNA is precious, so the aleurone cells make a disposable RNA copy of the gene that is turned on. This disposable copy of the information, a kind of blueprint, is often called messenger RNA. The process of making this RNA copy is called transcription. The RNA that was made in the transcription process is transported into the cytoplasm of the aleurone cells.

In the cytoplasm, the messenger RNA joins up with a ribosome to begin the process of making a protein. This process is often called protein synthesis or translation. In this process, the ribosome examines the information held in the sequence of bases in the RNA. Transfer RNAs charged with particular amino acids are moved into the position specified by the instructions in the messenger RNA, and the amino acids are joined in a proper sequence by the ribosome. The sequence of amino acids determines the properties of the protein being assembled.

In this case, the critical protein made with the information held in the RNA is amylase. This protein turns out to be an enzyme of great importance. The process of information held in the genes of DNA being transcribed into RNA and then translated into protein constitutes the central dogma of genetics. *A question one might ask right away is from where do the amino acid building blocks for the amylase come?*

The answer is from some other biochemistry in the aleurone cells. This biochemistry causes the storage proteins in the aleurone cells to be digested by hydrolytic enzymes. The hydrolysis is accelerated by enzymes known as proteases. These enzymes increase the rate at which the storage protein is cut into individual amino acids.

The amino acids released by the hydrolysis are then free to be reassembled by the ribosomes into the structure of amylase. The same thing happens in people. In a similar way, barley aleurone storage proteins are digested and amylase is made from the released amino acids. Amylase is not just any old protein. It happens to be an enzyme. It catalyzes (accelerates) a particular chemical reaction. Specifically, the amylase speeds the hydrolysis of starch into its component sugar units. Hurrah! The physiologists now know how the sugar is made in a seed. They can also tell the brewers how to increase the sugar yields for beer making.

Note that Gibberellic Acid was the first plant hormone purified in large quantities to be successfully used to increase the yield of any economically-important biochemical in plants! Remember that we wanted to get the seed germinated (sprouted). This does not occur until after the released sugar is transported from the endosperm to the embryo. The cotyledon is a major sugar transfer area in grains. Active sugar uptake is its specialized function. The sugar is transported into the embryo and used as a fuel and building block for growth of the embryo itself. The resulting embryo growth will include the emergence of the radicle from the seed/fruit coat.

Our barley seed story is great, but it does not explain some other forms of seed germination. The next great finding in the study of seed germination is what happens in lettuce seeds (*Lactuca sativa*). Lettuce seeds look somewhat different from barley seeds, because they are dicots. They also have separate seed coat and fruit walls. But the things we buy as lettuce "seeds" are really lettuce fruits (called cypselas). As different as they are, the seed germination process in lettuce needs just one step more than barley to get its seeds germinating.

Lettuce cypselas have very thin fruit walls and seed coats, so light can penetrate these coverings. As is typical of small seeds, light is needed to stimulate seed germination. One might wonder how this would fit into what we just found out with barley. The DNA must first be photoactivated before GA can have its effect. This is accomplished by the formation of a chemical signal known as Pfr. Pfr is one of two possible chemical forms of a molecule known as phytochrome. Phytochrome is a pigment that absorbs light energy. When it does absorb light energy its internal chemical structure is altered instantly.

Thus Pfr can be converted to the alternate Pr form of phytochrome by simply hitting the seeds with a flash of far-red light (730 nm). Similarly, the Pr form of phytochrome can be converted back into Pfr with a flash of red light (660 nm). Dry lettuce seeds have slightly more than half of their phytochrome in the Pfr form.

So if they are moistened and kept in the dark, some of the seeds will actually germinate. If we shine red (or it turns out even white) light on imbibing seeds, then virtually all of the phytochrome is converted to the Pfr form and this photoactivates the genes in the DNA. Almost all of the seeds will germinate in this light!

On the other hand, if we shine far-red light on imbibing seeds, then virtually all of the phytochrome in the seeds is converted to the Pr form and none of the seeds will germinate. Thus phytochrome and light tell lettuce seeds when they are exposed to sunlight and can begin to germinate. Small seeds, such as those in lettuce, do not have reserves enough to start germinating without being able to quickly get to light. It is no surprise that small-seeded species have evolved this mechanism for photoactivation of seed germination. It prevents a deeply-buried seed from trying to germinate.

SEED PRODUCTION

Seeds are produced in several related groups of plants, and their manner of production distinguishes the angiosperms ("enclosed seeds") from the gymnosperms ("naked seeds"). Angiosperm seeds are produced in a hard or fleshy (or with layers of both) structure called a fruit that encloses the seeds, hence the name. In gymnosperms, no special structure develops to enclose the seeds, which begin their development "naked" on the bracts of cones. However, the seeds do become covered by the cone scales as they develop in some species of conifer.

KINDS OF SEEDS

There are a number of modifications to seeds by different groups of plants. One example is that of the so-called *stone* fruits (such as the peach), where a hardened fruit layer (the endocarp) surrounds the actual seed and is fused to it. Many structures commonly referred to as "seeds" are actually dry fruits. Sunflower seeds are sold commercially while still enclosed within the hard wall of the fruit, which must be split open to reach the seed.

Seed Development

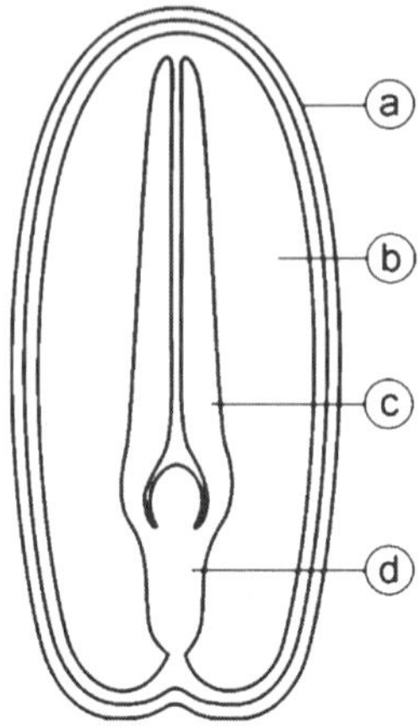

Fig. Diagram of the Internal Structure of a Dicot seed and Embryo. (a) Seed coat, (b) Endosperm, (c) Cotyledon, (d) Hypocotyl.

The inside of a *Ginkgo* seed, showing a well-developed embryo, nutritive tissue (megagametophyte), and a bit of the surrounding seed coat. The seed, which is an embryo with two points of growth (one of which forms the stems the other the roots) is enclosed in a seed coat with some food reserves. Angiosperm seeds consist of three genetically distinct constituents

- The embryo formed from the zygote,
- The endosperm, which is normally triploid,
- The seed coat from tissue derived from the maternal tissue of the ovule.

In angiosperms, the process of seed development begins with double fertilization and involves the fusion of the egg and sperm nuclei into a zygote.

The second part of this process is the fusion of the polar nuclei with a second sperm cell nucleus, thus forming a primary endosperm. Right after fertilization the zygote is mostly inactive but the primary endosperm divides rapidly to form the endosperm tissue. This tissue becomes the food that the young plant will consume until the roots have developed after germination or it develops into a hard seed coat.

The seed coat forms from the two integuments or outer layers of cells of the ovule, which derive from tissue from the mother plant, the inner integument forms the tegmen and the outer forms the testa. When the seed coat forms from only one layer it is also called the testa, though not all such testa are homologous from one species to the next.

In gymnosperms, the two sperm cells transferred from the pollen do not develop seed by double fertilization but instead only one sperm fertilizes the egg while the other is not used. The seed is composed of the embryo (the result of fertilization) and tissue from the mother plant, which also form a cone around the seed in coniferous plants like Pine and Spruce. The ovules after fertilization develop into the seeds; the main parts of the ovule are the funicle; which attaches the ovule to the placenta, the nucellus; the main region of the ovule were the embryo sac develops, the micropyle; A small pore or opening in the ovule where the pollen tube usually enters during the process of fertilization, and the chalaza; the base of the ovule opposite the micropyle, where integument and nucellus are joined together.

The shape of the ovules as they develop often affects the finale shape of the seeds. Plants generally produce ovules of four shapes the most common shape is called anatropous, with a curved shape. Orthotropous ovules are straight with all the parts of the ovule lined up in a long row producing an uncurved seed. Campylotropous ovules have a curved embryo sac often giving the seed a tight "c" shape. The last ovule shape is called amphitropous, where the ovule is partly inverted and turned back 90 degrees on its stalk or funicle.

In the majority of flowering plants the zygotes first division is transversely orientated in regards to the long axis and this establishes the polarity of the embryo. The upper or chalazal pole becomes the main area of growth of the embryo, while the lower or micropylar pole produces the stalk-like suspensor that attaches to the micropyle. The suspensor absorbs and manufacturers nutrients from the endosperm that are utilized during the embryos growth.

The embryo is composed of different parts; the epicotyle will grow into the shoot, the radicle grows into the primary root, the hypocotyl connects the epicotyle and the radicle, the cotyledons form the seed leaves, the testa or seed coat forms the outer covering of the seed. Monocotyledonous plants like corn, have other structures; instead of the hypocotyle-epicotyle, it has a coleoptile that forms the first leaf and connects to the coleorhiza that connects to the primary root and adventitious roots form from the sides.

The seeds of corn are constructed with these structures; pericarp, scutellum (single large cotyledon) that absorbs nutrients from the endosperm, endosperm, plumule, radicle, coleoptile and coleorhiza - these last two structures are sheath-like and enclose the plumule and radicle, acting as a protective covering. The testa or seed coats of both monocots and dicots are often marked with patterns and textured markings, or have wings or tufts of hair.

SEED SIZE AND SEED SET

Seeds are very diverse in size. The dust-like orchid seeds are the smallest with about one million seeds per gram. Embryotic seeds have immature embryos and no significant energy reserves. They are myco-heterotrophs, depending on mycorrhizal fungi for nutrition during germination and the early growth of the seedling, in fact some terrestrial Orchid seedlings spend the first few years of their life deriving energy from the fungus and do not produce green leaves. At over 20 kg, the largest seed is the coco de mer.

Plants that produce smaller seeds can generate many more seeds while plants with larger seeds invest more resources into those seeds and normally produce fewer seeds. Small seeds are quicker to ripen and can be dispersed sooner, so fall blooming plants often have small seeds. Many annual plants produce great quantities of smaller seeds; this helps to ensure that at least a few will end in a favourable place for growth. Herbaceous perennials and woody plants often have larger seeds, they can produce seeds over many years, and larger seeds have more energy reserves for germination and seedling growth and produce larger, more established seedlings.

SEVERAL FUNCTIONS OF SEEDS

Seeds serve several functions for the plants that produce them. Key among these functions are nourishment of the embryo, dispersal to a new location, and dormancy during unfavourable conditions. Seeds fundamentally are a means of reproduction and most seeds are the product of sexual reproduction which produces a remixing of genetic material and phenotype variability that natural selection acts on.

EMBRYO NOURISHMENT

Seeds protect and nourish the embryo or baby plant. Seeds usually give a seedling a faster start than a sporling from a spore gets because of the larger food reserves in the seed.

Seed Dispersal

Unlike animals, plants are limited in their ability to seek out favourable conditions for life and growth. As a consequence, plants have evolved many ways to disperse their offspring by dispersing their seeds. A seed must somehow "arrive" at a location and be there at a time favourable for germination and growth. When the fruits open and release their seeds in a regular way, it is called dehiscent, which is often distinctive for related groups of plants, these fruits include; Capsules, follicles, legumes, silicles and siliques. When fruits do not open and release their seeds in a regular fashion they are called indehiscent, which include these fruits; Achenes, caryopsis, nuts, samaras, and utricles.

Seed dispersal is seen most obviously in fruits; however many seeds aid in their own dispersal. Some kinds of seeds are dispersed while still inside a fruit or cone, which later opens or disintegrates to release the seeds. Other seeds are expelled or released from the fruit prior to dispersal. Milkweeds produce a fruit type, known as a *follicle,* that splits open along one side to release the seeds. Iris capsules split into three "valves" to release their seeds.

By Wind (Anemochory)

Dandelion seeds (achenes) can be carried long distances by the wind:

- Many seeds (*e.g.* maple, pine) have a wing that aids in wind dispersal.

- The dustlike seeds of orchids are carried efficiently by the wind.
- Some seeds, (*e.g.* dandelion, milkweed, poplar) have hairs that aid in wind dispersal.

By Water (hydrochory)

- Some plants, such as *Mucuna* and *Dioclea*, produce buoyant seeds termed sea-beans or drift seeds because they float in rivers to the oceans and wash up on beaches.

By Animals (Zoochory)

- Seeds (burrs) with barbs or hooks (*e.g.* acaena, burdock, dock which attach to animal fur or feathers, and then drop off later.
- Seeds with a fleshy covering (*e.g.* apple, cherry, juniper) are eaten by animals (birds, mammals) which then disperse these seeds in their droppings.
- Seeds (nuts) which are an attractive long-term storable food resource for animals (*e.g.* acorns, hazelnut, walnut); the seeds are stored some distance from the parent plant, and some escape being eaten if the animal forgets them.

Myrmecochory

Myrmecochory is the dispersal of seeds by ants. Foraging ants disperse seeds which have appendages called elaiosomes (*e.g.* bloodroot, trilliums, Acacias, and many species of Proteaceae). Elaiosomes are soft, fleshy structures that contain nutrients for animals that eat them. The ants carry such seeds back to their nest, where the elaiosomes are eaten.

The remainder of the seed, which is hard and inedible to the ants, then germinates either within the nest or at a removal site where the seed has been discarded by the ants. This dispersal relationship is an example of mutualism, since the plants depend upon the ants to disperse seeds, while the ants depend upon the plants seeds for food. As a result, a drop in numbers of one partner can reduce success of the other.

In South Africa, the Argentine ant (*Linepithema humile*) has invaded and displaced native species of ants. Unlike the native ant species, Argentine ants do not collect the seeds of *Mimetes cucullatus* or eat the elaiosomes. In areas where these ants have invaded, the numbers of *Mimetes* seedlings have dropped.

SEED DORMANCY AND PROTECTION

One important function of most seeds is delaying germination, which allows time for dispersal and prevents germination of all the seeds at one time when conditions appear favourable. The staggering of germination safeguards some seeds or seedlings from suffering during short periods of bad weather, transient herbivores or competition from other plants for light and nutrients. Many species of plants have seeds that germinate over many months or years, and some seeds can remain in the soil seed bank for more than 50 years before germination.

Seed dormancy is defined as a seed failing to germinate under environmental conditions optimal for germination, normally when the seed's environment is at the right temperature with proper soil moisture conditions. Induced dormancy or seed quiescence occurs when a seed fails to germinate because the external environmental conditions are inappropriate for germination, mostly in response to being too cold or hot, or too dry.

True dormancy or innate dormancy is caused by conditions within the seed that prevent germination under normally ideal conditions. Often seed dormancy is divided into four major categories exogenous; endogenous; combinational; and secondary. Exogenous dormancy is caused by conditions outside the embryo including

- Hard seed coats or physical dormancy occurs when seeds are impermeable to water or the exchange of gases. In some seeds the seed coat physically prevents the seedling from growing.
- Çhemical dormancy includes growth regulators etc.

Endogenous dormancy is caused by conditions within the embryo itself, including:

- Immature embryos where some plants release their seeds before the tissues of the embryos have fully differentiated, and the seeds ripen after they take in water while on the ground, germination can be delayed from a few weeks to a few months.
- Morphological dormancy where seeds have fully differentiated embryos that need to grow more before seed germination, the embryos are not yet fully developed.
- Morphophysiological dormancy seeds with underdeveloped embryos, and in addition have physiological components to dormancy. These seeds therefore require a dormancy-breaking treatments as well as a period of time to develop fully grown embryos.
- Physiological dormancy prevents seed germination until the chemical inhibitors are broken down or are no longer produced by the seed, often physiological dormancy is broken by a period of cool moist conditions, normally below (+4°C) 39°F, or in the case of many species in *Ranunculaceae* and a few others, (-5°C) 24°F. Other chemicals that prevent germination are washed out of the seeds by rainwater or snow melt. Abscisic acid is usually the growth inhibitor in seeds and its production can be affected by light. Some plants like Peony species have multiple types of physiological dormancy, one affects radical growth while the other affects shoot growth.
  - Drying; some plants including a number of grasses and those from seasonally arid regions need a period of drying before they

will germinate, the seeds are released but need to have a lower moister content before germination can begin. If the seeds remain moist after dispersal, germination can be delayed for many months or even years. Many herbaceous plants from temperate climate zones have physiological dormancy that disappears with drying of the seeds. Other species will germinate after dispersal only under very narrow temperature ranges, but as the seeds dry they are able to germinate over a wider temperature range.

— Photodormancy or light sensitivity affects germination of some seeds. These photoblastic seeds need a period of darkness or light to germinate. In species with thin seed coats, light may be able to penetrate into the dormant embryo. The presence of light or the absence of light may trigger the germination process, inhibiting germination in some seeds buried too deeply or in others not buried in the soil.

— Thermodormancy is seed sensitivity to heat or cold. Some seeds including cocklebur and amaranth germinate only at high temperatures (30°C or 86°F) many plants that have seed that germinate in early to mid summer have thermodormancy and germinate only when the soil temperature is warm. Other seeds need cool soils to germinate, while others like celery are inhibited when soil temperatures are too warm. Often thermodormancy requirements disappear as the seed ages or dries.

Combinational dormancy also called double dormancy. Many seeds have more than one type of dormancy, some *Iris* species have both hard impermeable seeds coats and physiological dormancy. Secondary dormancy is caused by conditions after the seed has been dispersed and occurs in some seeds when none dormant seed is exposed to conditions that are not favourable to germination, very often high temperatures.

The mechanisms of secondary dormancy is not yet fully understood but might involve the lose of sensitivity in receptors in the plasma membrane. Many garden plants have seeds that will germinate readily as soon as they have water and are warm enough, though their wild ancestors may have had dormancy, these cultivated plants lack seed dormancy. After many generations of selective pressure by plant breeders and gardeners dormancy has been selected out.

For annuals, seeds are a way for the species to survive dry or cold seasons. Ephemeral plants are usually annuals that can go from seed to seed in as few as six weeks. Not all seeds undergo a period of dormancy. Seeds of some mangroves are viviparous, they begin to germinate while still attached to the parent. The large, heavy root allows the seed to penetrate into the ground when it falls.

SUPPLY OF STORED ENERGY OF SEED STRUCTURE

A seed contains the embryo from which a new plant will grow under proper conditions. Seeds also usually contain a supply of stored energy and is wrapped in the seed coat or testa. Seeds are very diverse in size. The dust-like orchid seeds are the smallest with about one million seeds per gram. Embryotic seeds have immature embryos and no significant energy reserves. They are myco-heterotrophs, depending on mycorrhizal fungi for nutrition during germination. At over 20 kg, the largest seed is the coco de mer.

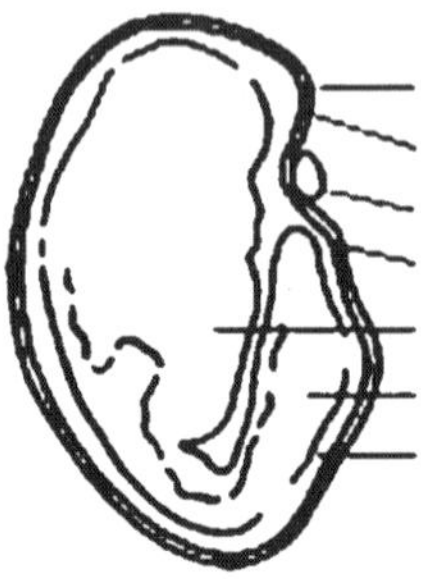

Fig. Alfalfa Seed Structure

The embryo has one cotyledon or seed leaf in monocotyledons, two cotyledons in almost all dicotyledons and two or more in gymnosperms. The radicle is the embryonic root. The plumule is the embryonic shoot. The embryonic stem above the point of attachment of the cotyledon(s) is the epicotyl. The embryonic stem below the point of attachment is the hypocotyl. In angiosperms, the stored food begins as a tissue called the endosperm, which is derived from the parent plant via double fertilization. The usually triploid endosperm is rich in oil or starch and protein. In gymnosperms, such as conifers, the food storage tissue is part of the female gametophyte, a haploid tissue. In some species, the embryo is embedded in the endosperm or female gametophyte, which the seedling will use upon germination.

In others, the endosperm is absorbed by the embryo as the latter grows within the developing seed, and the cotyledons of the embryo become filled with this stored food. At maturity, seeds of these species have no endosperm and are termed exalbuminous seeds. Some exalbuminous seeds are bean, pea, oak, walnut, squash, sunflower, and radish. Seeds with an endosperm at maturity are termed albuminous seeds. Most monocots (*e.g.* grasses and palms) and many dicots (*e.g.* brazil nut and castor bean) have albuminous seeds. All gymnosperm seeds are albuminous. The seed coat develops from the tissue, the integument, originally surrounding the ovule.

The seeds of angiosperms are contained in a hard or fleshy (or with layers of both) structure called a fruit. Gymnosperm seeds begin their development "naked" on the bracts of cones, although the seeds do become covered by the cone scales as they develop. An example of a hard fruit layer surrounding the actual seed is that of the so-called *stone* fruits (such as the peach). Some seeds have an appendage on the seed coat such an aril (as in yew and nutmeg) or an elaiosome (as in Corydalis) or hairs (as in cotton). The hilum is the scar on the seed coat where the seed was attached to the ovary wall by the funiculus.

In order for the seed coat to split, the embryo must imbibe (soak up water), which causes it to swell, splitting the seed coat. However, the nature of the seed coat determines how rapidly water can penetrate and subsequently initiate germination. For seeds with a very thick coat, scarification of the seed coat may be necessary before water can reach the embryo.

Examples of scarification include gnawing by animals, freezing and thawing, battering on rocks in a stream bed, or passing through an animal's digestive tract. In the latter case, the seed coat protects the seed from digestion, while perhaps weakening the seed coat such that the embryo is ready to sprout when it gets deposited (along with a bit of fertilizer) far from the parent plant. In species with thin seed coats, light may be able to penetrate into the dormant embryo.

The presence of light or the absence of light may trigger the germination process, inhibiting germination in some seeds buried too deeply or in others not buried in the soil. Abscisic acid is usually the growth inhibitor in seeds. A fertilized seed contains the embryo from which a new plant will grow under proper conditions. It also contains a supply of stored food and is wrapped in the seed coat or testa. The stored food begins as a tissue called endosperm derived from the parent plant. Endosperm becomes rich in oil or starch, and protein. In some species, the embryo is imbedded in the endosperm, which the seedling will use upon germination. In others, the endosperm is absorbed by the embryo as the latter grows within the developing seed.

A fertilized seed contains the embryo from which a new plant will grow under proper conditions. It also contains a supply of stored food and is wrapped in the seed coat or *testa*. The stored food begins as a tissue called *endosperm* derived from the parent plant. Endosperm becomes rich in oil or starch, and protein.

In some species, the embryo is imbedded in the *endosperm,* which the seedling will use upon germination. In others, the endosperm is absorbed by the embryo as the latter grows within the developing seed, and the cotyledons of the embryo become filled with this stored food. At maturity, seeds of these species have no endosperm. Some common plant seeds that lack an endosperm are bean, pea, oak, walnut, squash, sunflower, and radish. Plant seeds with an endosperm include all conifers and most monocotyledons (*e.g.* grasses and palms), and also *e.g.* brazil nut, castor bean.

The seed coat develops from tissues (called *integument*) originally surrounding the ovule. The seed coat in the mature seed can be a paper thin layer or something more substantial. The seed coat helps protect the embryo from mechanical injury and from drying out. In order for the seed coat to split, the embryo must imbibe (soak up water) which causes it to swell, splitting the seed coat.

However, the nature of the seed coat determines how rapidly water can penetrate and subsequently initiate germination. For seeds with a very thick coat, scarification of the seed coat may be necessary before water can reach the embryo. Examples of scarification include gnawing by animals, freezing and thawing, battering on rocks in a stream bed, or passing through an animal's digestive tract.

In the latter case, the seed coat protects the seed from digestion, while perhaps weakening the seed coat such that the embryo is ready to sprout when it gets deposited (along with a bit of fertilizer) far from the parent plant. In species with thin seed coats, light may be able to penetrate into the dormant embryo. The presence of light or the absence of light may trigger the germination process, inhibiting germination in some seeds buried too deeply or in others not buried in the soil. Abscisic acid is usually the growth inhibitor in seeds.

The seeds of angiosperms are contained in a hard or fleshy (or with layers of both) structure called a fruit. Gymnosperm seeds begin their development "naked" on the bracts of cones, although the seeds do become covered by the cone scales as they develop. An example of a hard fruit layer surrounding the actual seed is that of the so-called *stone* fruits (such as the peach).

Alfalfa is a dicotyledonous plant. That is, the seed is composed of two embryonic or seed leaves called cotyledons. In addition, the seed contains the primary root or radicle, the shoot growing point or epicotyl located above the cotyledons, and the endosperm or food storage area. Externally the seed has several visible structures – the hilum or point of attachment in the seed pod, the lens or a weak point in the seed coat, and the micropyle, which is a remnant of the tiny opening the pollen tube grew through during the process of fertilizing the female ovary.

Alfalfa Seed Structure

Germination is the resumption of growth by the embryo within the seed leading to the development of a new plant. The germination process is complete as soon as the radicle ruptures the seed coat.

The process is influenced by available soil moisture, soil temperature, and the nature of the soil surrounding the seed (*i.e.*, salty vs. normal soil) or residual herbicides that may be present in the soil.

Alfalfa Germination Process

Seedling growth is the developmental period of the young plant from the time germination is completed until it can manufacture enough food through photosynthesis to sustain growth. The seedling root is the first structure to emerge from the seed during germination. It penetrates the soil very rapidly, forming a slender, usually unbranched taproot, which may penetrate 5 to 6 feet into the soil during the first growing season. Once the seedling root is anchored firmly in the soil, the seedling axis below the cotyledons elongates in an arch (hypocotyl arch) pulling the cotyledons upward to the soil surface.

Seed germination and seedling emergence occur in about three to seven days. Fewer days may be required for seedling emergence under ideal soil moisture and temperature conditions. As the hypocotyl arch emerges from the soil, growth stops on the side exposed to light and continues on the underside until the seedling is in an upright position. This raises the cotyledons above the soil surface where they expand. The growing point (epicotyl) of the seedling is now exposed. The first true leaf is unifoliolate and emerges from a bud at the first stem node above the cotyledons. Seedling growth is complete. Under good growing conditions, the seedling is developed fully 10 to 15 days after planting.

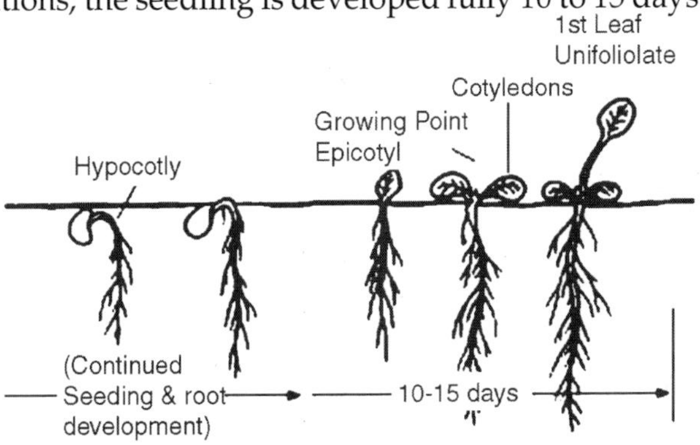

Fig. Seedling Emergence Through Unifoliolate Leaf Stage

Alfalfa Seedling Growth

A fully developed alfalfa seedling does not assure plant establishment. It must continue to develop deeper roots and grow more leaves to survive and become an established stand. Vegetative growth continues through cell division and expansion in the epicotyl or growing point of the young plant.

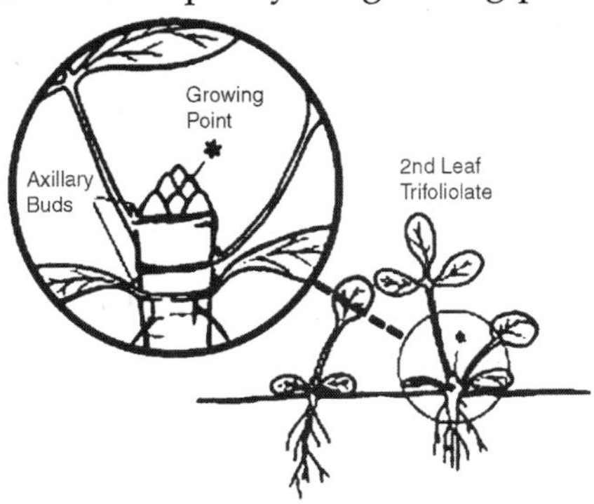

Fig. Unifoliolate Through Fourth Trifoliolate Leaf Stage

The second leaf of the alfalfa plant is usually trifoliolate (three leaflets) and originates from the second primary stem node. All subsequent leaves are trifoliolates except in new multifoliolate varieties that have 5, 7, or 9 leaflets per leaf. The unifoliolate leaf and all subsequent trifoliolate/multifoliolate leaves are useful characters to distinguish alfalfa from weed seedlings in stand evaluation. Axillary buds develop in the axils of all leaves. After three or more trifoliolate/multifoliolate leaves have appeared on the primary stem, new secondary stem growth may occur from any of the axillary buds, but frequently only one axillary bud stem (usually the unifoliolate) develops early, especially if seedlings are shaded in a companion crop or under competition from weed growth or other alfalfa plants.

Alfalfa First Trifoliolate Leaf and Buds

The primary and secondary stem(s) of the young plant increase in length by cell division and internode elongation from the first stem node upwards. The second and subsequent leaves are trifoliolate/multifoliolate and develop alternately at each stem node as growth continues.

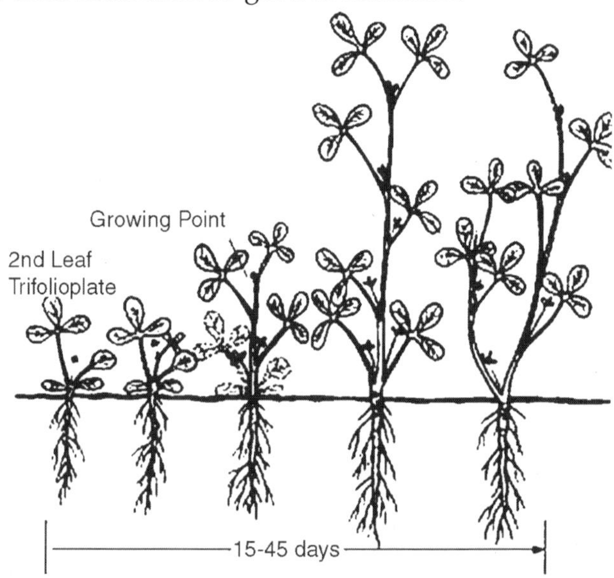

Fig. First-year Vegetative Growth and Development

Once the first true trifoliolate leaf develops, further growth and development is best described by the number of trifoliolate/multifoliolate leaves that develop on the main shoot as the plant continues to grow. Any axillary buds in axils of leaves can develop new stem tissue and generally do in less competitive conditions.

Alfalfa First-Year Vegetative Development

Growth and development of new shoots from axillary buds gives the young plant a branched appearance, especially if light is adequate and the stand is not too thick. Vigourously growing alfalfa plants quite often have three and sometimes four secondary shoots in addition to the primary or central stem, which forms the characteristic first-year crown.

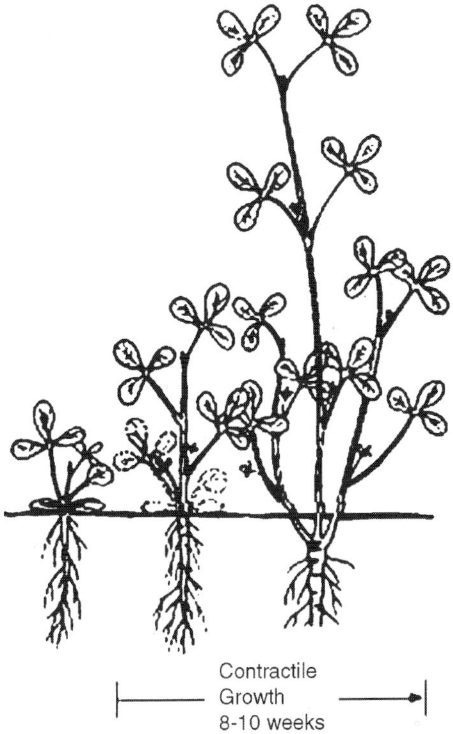

Fig. Crown Development Showing Contractile Growth

The branches from the unifoliolate leaf bud appears first, followed by branches from the cotyledonary and first trifoliolate leaf buds. Fewer branches may form the crown with late seedings, under severe companion crop competition, or with high seeding rates. Under these conditions the cotyledonary, unifoliolate, and first trifoliolate leaf buds may remain dormant or may give rise to a branch rhizome or underground stem in the fall.

The young alfalfa plant, at about 8 to 10 weeks of age, undergoes a growth phase known as contractile growth. This process in alfalfa and sweetclover involves a change in the shape of cells in the hypocotyl or seedling axis below the cotyledons and upper portion of the primary root from long and narrow to short and wide as a result of carbohydrate or food storage. This shift pulls the lower stem nodes beneath the soil surface.

Most winter hardy alfalfa varieties have several nodes pulled well below the soil surface (1 to 3 inches) in the seeding year. Contractile growth greatly aids winter survival of alfalfa by providing soil insulation for the perennial over-wintering crown structures.

Crown Development

The lower internodes or nodes of the primary or secondary stems may produce adventitious or crown buds. Crown buds or dominant axillary buds produce the vegetative regrowth following harvest the seedling year, or these buds may give rise to branch rhizomes in the fall. The branch rhizome is the structure from which new growth will initiate in the spring unless winter injury occurs. The branch rhizome must be developed adequately in the fall by the first killing frost or the seeding will fail, even under the most ideal growing conditions. In most years, alfalfa seeded later than about August 15 under good moisture (irrigation) conditions in North Dakota will not have adequate time to initiate branch rhizomes before the first killing frost. A dryland seeding after the first week of August is not recommended because of the increased risk of stand loss.

A crown of a perennial can be described best as the young overwintering stem tissue. The crown in the fall of the first year may be as small as the lower portion of the main stem and dormant cotyledonary node buds, or as large as the lower portion of the main stem (*i.e.*, colytedonary, unifoliolate, and first trifoliolate leaf nodes), the secondary stems from these nodes, and the branch rhizomes that develop.

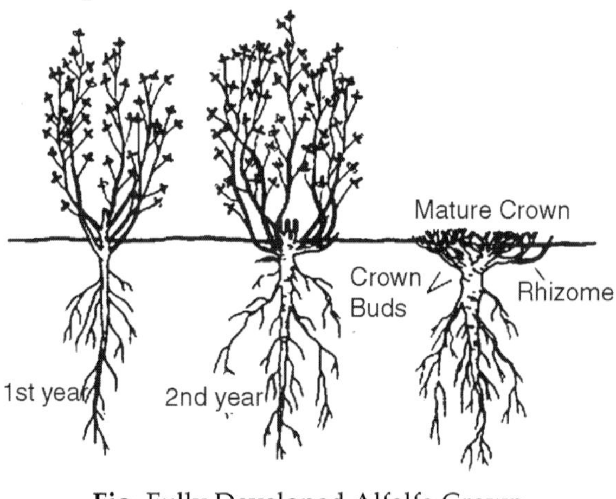

Fig. Fully Developed Alfalfa Crown

Spring growth will occur from the rudimentary leaves on the branch rhizome if winter injury hasn't "burnt" them off. If winter injury has occurred, spring growth will initiate from adventitious buds on the branch rhizome or from a dormant crown bud. But, spring growth is much slower in this case. All subsequent growth in the first-harvest year occurs from adventitious or dormant crown buds.

The crown of the alfalfa plant increases in size during the second year. Branch rhizomes develop on last year's branch rhizome. These grow outward and upward increasing the circumference of the plant. After two to four years, the typical multibranched crown of alfalfa develops.

Alfalfa Wintering Structures

The crown and associated root tissue are the storage centre for carbohydrates (food) produced through the process of photosynthesis. The carbohydrates are used to develop cold resistance or winter hardiness, for plant respiration during winter, and to provide the energy source for the initiation of the new growth in the spring and following each harvest.

First-year management influences subsequent growth. Alfalfa management is discussed in R-571, Alfalfa Management in North Dakota. Over the next several lectures we will follow the early growth of young trees and other vascular plants from seed through germination and growth in the first year.

We will draw examples from many species, but will illustrate some phases of growth by contrasting two tree species native to British Columbia Douglas-fir representing gymnosperms, and red alder representing angiosperms.

Evolution of Seeds

The evolution of seeds was a critical factor in facilitating the proliferation of land plants into a wide range of environments. Seeds protect and nourish embryos, and embryos can survive a wide range of environmental conditions Seeds can also persist in a dormant state, enabling plant species to survive and regenerate following catastrophic events or periods of weather that are not conducive for growth.

Seeds can remain viable for long periods of time, centuries, even millenia. An extreme example is provided by frozen 10,000 year old lupine seeds found in the Yukon. These seeds germinated within 48 hours of thawing!

Gymnosperms

Gymnosperm seeds are, by definition, naked, that is they lack a surrounding fruit. The outermost layer of the seed is the seed coat. It originates from the mother tree and is diploid. The seed coat has three layers the outer layer; the thicker, tough 'stony' or middle layer; and the inner layer. The colour of the seed coat, which can be quite variable, is due to tannins that accumulate during seed development. Some species of conifers, including *Abies* spp. (true firs), *Tsuga* spp. (hemlocks) and *Thuja plicata* (western redcedar) have resin vesicles in the middle or outer layers of the seed coat.

These resin vesicles may play a role in:

- Seed coat dormancy,
- Protecting the embryo from excessive dehydration,
- Deterring seed herbivory.

These vesicles can also create problems for the extraction and cleaning of conifer seed. Many conifer seeds have a seed wing to aid in dispersal. Immediately inside the cell wall is the nucellus, a papery layer surrounding the megaspore cell wall. Inside the megaspore cell wall is the megagametophyte (the female gametophyte), the haploid nutritional tissue found in gymnosperm seeds. Early in the development of cones, megagametophytes produce egg cells which are fertilized by male gametes to produce zygotes. Zygotes develop into embryos. The megagametophyte then plays its second functional role, surrounding the embryo, protecting and nourishing it.

The megagametophyte in Douglas-fir is 60 per cent lipids, 16 per cent proteins, and 2 per cent sugars, making it a high-energy and nutrient tissue both for the embryos it contains and for a plethora of seed predators including small mammals, birds and insects. In a dry seed (<10 per cent moisture), the megagametophyte is creamy yellow and does not entirely fill the seed coat. When a seed is imbibed (soaked in water until the moisture content is 30-40 per cent), the megagametophyte is white and shiny.

The embryo is found in the corrosion cavity, a pit in the centre of the megagametophyte that is fully filled by the embryo in mature seeds. It consists of the cotyledons, shoot apical meristem, root apical meristem, root cap and suspensor. The cotyledons and shoot apical meristem point towards the wider chalazal end of the seed.

The radicle (embryonic root including the apical meristem and root cap) and suspensor are at the more pointed micropylar end of the seed towards the micropyle (the opening in the integument through which the pollen entered the ovule). The number of cotyledons varies from 2 in western redcedar to 12 or more in *Pinus ponderosa* (ponderosa pine). The suspensor is found at the base of the root cap and plays a role early in embryo development by pushing the embryo into the megagametophyte.

Angiosperms

Angiosperm seeds are found inside fruit. The many types of fruit will be discussed in detail in lab. Angiosperm seeds generally vary more in structure than conifer seeds, but share some features with conifer seeds and have different parts playing the same functional roles for others. Like conifer seeds, angiosperm seeds are surrounded by a seed coat protecting the embryo. Inside the seed coat (which is usually formed by the integument, but is sometimes fused with the fruit), nutritive reserves for the mature embryo may be stored either in endosperm or in large cotyledons.

Endosperm is formed through the fertilization of the two polar nuclei by a male gamete, producing triploid cells, at the same time that the egg cell is fertilized by a second male gamete (more on this after the Christmas break). All angiosperm seed have endosperm early in embryo development. In mature seeds of some species, there may be a substantial amount of triploid endosperm surrounding the embryo (*e.g., Arbutus menziesii* (arbutus)). Commonly in Eudicots, however, embryos largely absorb the endosperm and the nutritional reserves are stored in the two (of course!) large cotyledons that fill the seed cavity (*e.g., Alnus* spp. (alders), *Acer* spp. (maples), and *Quercus* spp. (oaks)). Some angiosperm seeds already contain a primordial shoot with leaves, called a plumule, above (distal to) the cotyledons prior to germination.

The seeds of monocots have several noteworthy features. They typically have endosperm at maturity, and the single cotyledon often plays a role in absorbing food digested from the endosperm. Grass embryos have a large cotyledon called a scutellum attached to one side, as well as a plumule and a radicle. The plumule is protected by a structure called the coleoptile, while the radicle is protected by the coleorhiza.

Angiosperm seed also have suspensors, but these have a different function than those of conifers. In angiosperms, suspensors are involved in supporting the early development of the embryo both through providing nutrition and growth regulators. The suspensor cells die in angiosperms relatively early in embryo development.

SEED DISPERSAL

Conifers

The dispersal of winged seeds depends on seed mass, wing size and wind speed. Douglas-fir seeds have a mass of 9 to 17 mg, are 1 to 1.5 cm in length including wings, and can disperse up to 3 or 4 tree heights from the mother tree. Periodic weather events may result in rare dispersals up to a km in distance.

Most BC conifers shed seed August through October. Cone tissues die, cones dry out, cone scales flex and release seed. Dry conditions promote the flexing of cone scales and opening of cones, so cycles of wet and dry weather may produce repeated cycles of seed release. Seeds that land on snow may blow considerable distances along the frozen surface in winter.

Not all conifers have cones that open on maturity. The true firs (*Abies* spp.) bear upright rather than pendant cones, and have deciduous cone scales. When these cones are mature, the cone scales fall off the central cone rachis, releasing the seed. Whitebark pine (*Pinus albicaulis*) cones do not open and must be mechanically torn apart to release seeds. This is done by Clark's nutcrackers, which then disperse store seeds for later consumption (and regeneration of the pine).

Angiosperms

Dispersal of seeds in angiosperms can be via wind, through animal ingestion or through animal transport to food caches or as 'hitchhikers' (*e.g.,* burrs). The acorns of oaks, and the many fleshy fruits and tasty nuts produced by angiosperms, provide examples of angiosperms dispersed by animals.

Red alder is a wind-dispersed species with single seeds found in small (1.5 million per kg), winged nutlets. The fruit are shed in large numbers in mid to late fall, and can be wind-dispersed in air or across snow. Red alder is a good example of a pioneer species, capable of colonizing bare ground. Pioneer species typically have small seed produced in large numbers that disperse widely via wind.

Species of poplars and aspen (*Populus* spp.) are also pioneers, and produce large quantities of tiny seed (8 million per kg for trembling aspen). Poplar seed contain very small nutritional reserves and thus are quite short-lived. Fireweed (*Epilobium angustifolium*) is a good example of an herbaceous pioneer with seed production and dispersal similar to poplars.

SEED DORMANCY

Seed of most temperate and boreal species is dormant when it disperses, that is it won't germinate even if conditions of temperature, moisture and light are favourable. Why? To avoid active growth until winter has passed. An actively growing plant cannot be adequately frost hardy. Dormant seeds cease growth and maintain a very low metabolic rate until specific environmental signals are received or other requirements have been met.

Embryo Dormancy

Dormancy in many species is under the physiological control of the embryo. Embryo dormancy results from the levels of plant growth regulators (plant hormones). Dormant embryos have growth-inhibitors present and lack growth promoters. Breaking embryo dormancy is dependent on decreasing growth inhibitor concentrations and increasing growth promoters above some threshold.

Embryo dormancy in many species can be overcome through exposure to low temperatures to meet the chilling requirement of the seed. To accomplish this, seeds are imbibed (soaked in water), then placed in a fridge, usually on damp paper or moss. This is called stratification.

In Douglas-fir, holding seeds at 4°C for 4 weeks will meet the chilling requirement. Many species require a longer stratification period – up to 100 days. Seeds of some species with embryo dormancy may germinate eventually without chilling after being soaked in water, but will germinate more slowly than stratified seed, and the per cent of seeds that germinate will likely be lower. Embryo dormancy can also be overcome in some cases by artificially applying growth regulators. Some species require alternating day and night temperatures, or cycles of wetting and drying following low temperatures. Some species have both embryo dormancy and other types of seed dormancy.

Seed Coat Dormancy

Some desert and tropical species have chemical growth inhibitors in the seed coat that must be leached out by water before seed will germinate, increasing the probability that seed will germinate during adequately moist periods. Other species require a heat treatment to germinate. Germination of buckbrush (*Ceonothus* spp.) is promoted by pouring boiling water over seeds.

Other Types of Seed Dormancy

Embryo immaturity – Seeds of some species are immature at the time fruit or seed are shed and require 'after-ripening'. The seed of whitebark pine are typically immature when harvested by Clark's nutcrackers and continue to mature after caching.

Seed coat impermeability – Seed coats can be impermeable to water or oxygen, which physically imposes dormancy through preventing growth. This is common for leguminous species, among others. These seed may need scarification (scratching or abrading of the seed coat, either in nature or artificially with sandpaper), infection of seed coat by fungi or bacteria, or treatment with concentrated sulphuric acid (simulating passing the seed through the digestive tract of an animal).

Light requirement – Some species (e.g, some spruce and pine) require exposure to light to germinate. Other species require exposure to light of a specific quality. Red light promotes germination in these species, while far red light inhibits it. It is the ratio of far red to red light that is critical. As light passes through vegetation, this ratio increases, and germination in light-sensitive species will decrease. This means that seeds will be less likely to germinate when they are below competing vegetation. Once the competing vegetation is gone, the amount of red to far red light will increase, and the seed will germinate.

Serotiny – Most interior lodgepole pine (*Pinus contorta* ssp. *latifolia*) has cones that are serotinous. The cone scales are 'glued' closed by resin that has a melting point around 40°C. These cones must be exposed to either a fire or to particularly warm temperatures (*e.g.*, on a dark surface in the full sun) to open. The cones of juvenile interior lodgepole pine, and Shore pine (*P. contorta* ssp. *contorta*) do not have this adaptation – why?

Species Without Seed Dormancy

Red alder seeds lack dormancy, but since they are shed late in the fall, environmental conditions are not favourable for germination until spring. Other examples of seeds with no dormancy include sugar maple and willows, which are viable for only a few days following dispersal. Many agricultural crop species have no seed dormancy, but their wild progenitors do – humans have selected against seed dormancy in these species as it complicates farming.

Germination

Germination is the resumption of growth by embryos when exposed to favourable conditions. It requires several steps:

- Seeds imbibe and swell; the seed coat ruptures;

- Cell division occurs in the embryo and the radicle elongates and extends out of the seed;
- Further elongation of the radicle pushes either the cotyledons or just the epicotyl hormones produced in the meristems and possibly elsewhere initiate enzymatic activity, hydrolysis of stored nutrients, and movement of these nutrients to regions of growth.

The process of germination is complete when enough photosynthetic tissue is exposed to support further growth. The radicle develops into the tap root of the seedling (except in the case of species with fibrous root systems). It is positively gravitropic (grows in the direction of gravitational pull) while the shoot is negatively gravitropic (grows against gravity). The root cap consists of loose, slimy cells that protect the root apical meristem as the root elongates through the soil. The radicle develops root hairs for absorption of water and dissolved nutrients soon after germination.

In species with epigeous germination (*e.g.*, conifers, red alder, maples), the cotyledons are pushed out of the ground, turn green and become photosynthetically active. In species with hypogeous germination (*e.g.*, oaks and other species with large seeds), the cotyledons remain in the ground and only the epicotyl emerges. The energy reserves in the cotyledons supply the energy for early seedling growth, but the cotyledons never become photosynthetically active in this case. Rather, the first true leaves are the first photosynthetic organs with hypogeous germination.

Symbiotic relationships between seedling roots and other organisms are often initiated soon after germination. Lateral roots can become infected by mycorrhizal fungi soon after germination. In addition to becoming mycorhizal, red alder germinants becomes symbiotic with *Frankia* bacteria. Nodules form on the roots, and in these nodules N_2 gas is fixed into NH_4^+, which can then be converted into amino acids.

This allows red alder to 'fertilize itself'. Up to 320 kg N can be fixed per ha per year! Other species that fix nitrogen include *Ceonothus* spp. (deer brush) and *Shepherdia canadensis* (soopallalie).

7

Seed Dispersal by Animals

Plants, being stationary, require a mobile mode for seed dispersal. Modes for seed dispersal include self-projectile mechanisms, wind, water, and animals. Dispersal by animals falls into the category of plant-animal interactions, a subject of interest because of the reciprocal adaptations that can be observed. These reciprocal adaptations lead to coevolution, or change in organisms as a result of their interactions with each other.

ANIMALS AS DISPERSAL AGENTS

Plants that depend on animals for dispersal have seeds that are adapted to travelling on the outside or the inside of the animal. Seeds with burrs or hooks can attach to an animal's fur. For seeds that are transported internally, plants provide an attractive fruit pulp reward in return for the ride. The process of internal seed dispersal begins with the plant producing fruits. Next, an animal is attracted to these fruits based on their characteristics (such as odour), and begins to feed. After these fruit-eating animals (also known as frugivores) ingest fruit, they process them in their digestive tracts, and then regurgitate seeds or deposit seeds in their feces.

Behaviour, both innate and learned, is involved in every step of the dispersal process. First, the stimulus of being hungry makes the animal begin searching for food. While in this state of searching, the animal is receptive to stimuli (*e.g.*, colour, odour) that may signal the presence of food items. If the fruit encountered has an acceptable taste, the animal will probably choose to eat it. Where seeds are deposited depends on how long the animal stays at the plant feeding, and where the animal goes afterwards.

Not all animals that eat fruit are appropriate dispersers. The effectiveness of an animal as a seed dispersal agent depends on how many seeds it disperses, and how it treats these seeds. Animals that visit a plant for fruit more frequently may be more reliable than those animals that eat fruit more rarely. Some fruit-eating animals are considered seed 'predators' if they kill seeds by digesting the seed along with the fruit pulp. On the other hand, seeds can sprout better after passing through the digestive tract of certain frugivores. In addition to enhancing sprouting, an effective dispersal agent deposits seeds in appropriate habitats for their survival to reproductive adulthood. The survival of a seed greatly depends on where it lands. Seeds that move farther away from other seeds have greater success because they can better escape resource competition, interbreeding with parents, and post-dispersal mortality. Because not all animals that eat fruit are equally good at dispersing seeds, plants should change fruit characteristics to discourage frugivores that are seed predators, and encourage frugivores that are effective dispersers.

FRUIT CHARACTERISTICS AND FORAGER CHOICE

Fruit characteristics (or fruit traits) that foragers use to choose fruit as food items include toxin content, fruit appearance, and nutrient content. Together, the presentation of these traits is called the fruiting display. Foragers learn to recognize these traits as cues to trigger selection or avoidance of certain fruits as food items. The reliability of the behavioural response depends on the consequences of eating the fruit selected. Animals will prefer fruits that fulfil nutritional requirements, but avoid those that are toxic. Reciprocal change (coevolution) between plants and frugivores can occur when behaviours of foragers select fruit traits that in turn select for foraging behaviour.

Toxins: Some fruits contain toxins to deter seed pathogens or predators. Animals can learn which fruits are toxic, and thus learn to avoid them. Toxins can also limit the amount of time a more toxin-tolerant frugivore spends feeding at the plant. Because animals are limited to the amount of toxins they can handle over a period of time, a forager must stop eating a food item when maximum toxin load is reached. The foraging animal may then leave to find an alternative source of food or to seek antidotes. By forcing an animal to leave early, the plant may be able to ensure that its seeds will be deposited far enough away to avoid inbreeding, competition, and pathogens.

Fruit Appearance: Although toxin content limits ingestion of fruit, appearance can both limit and encourage it. Fruit appearance provides salient cues for recognition of fruit as a food source. Pigments such as carotenoids, flavonoids, and betalains give fruits their colours, and volatile compounds provide odors. In general, bird-dispersed fruit are red or black, and mammal-dispersed fruit have distinctive odors. The morphology (size and shape) of a fruit may also influence forager choice. Animals are limited to food items small enough to fit in their mouths and throats, especially birds and reptiles that tend to swallow fruits whole. In addition, birds may also consider the bulkiness of seeds when choosing fruit. Not only does seed bulk take up space in the digestive tract, it also may need to be jettisoned before flight.

Nutrient Content: Finally, fruit nutrients may be important in influencing forager choice. For most temperate fruits, energy in the form of sugars is the most abundant nutrient. Other vitamins and minerals may also be influential. For example, rose hips are known to be high in vitamin C. Nutrients may be more indirect in effect than appearance. Fruit appearance provides the cue for fruit choice, but nutrients provide the positive feedback that maintains an animal's preference for the fruit.

By understanding what influences an animal's decision to choose fruit, we can attempt to predict whether or not a plant's fruiting display will lead to successful dispersal. The fruiting display must balance maximizing gains from attracting effective dispersers while minimizing losses from also attracting ineffective dispersers. Thus, we can determine the effectiveness of a fruiting display by the number of seeds successfully dispersed per effort for the entire fruit crop. Effort can be determined by identifying the factor that influences an animal's decision to choose the fruit. For example, if the animal selects fruit on the basis of energy content, then effort is the amount of kilojoules of energy of the entire fruit crop. Effectiveness is then measured in units of numbers of seeds dispersed successfully relative to the amount of energy of the fruit crop.

In conclusion, foraging behaviour is an important component of the plant-seed disperser interaction. Foraging behaviour determines how fruits are chosen, and where the seeds are deposited. Toxin content, fruit appearance, and nutrient content may influence how fruits are selected and where seeds are deposited. By examining the behavioural components of the seed dispersal process, we can better understand how coevolution between plants and their dispersers may take place.

DISPERSAL BY ANIMALS

Snails disperse the small seeds of a very few plant species (*e.g.*, *Adoxa*). Earthworms are more important as seed dispersers. Many intact fruits and seeds can serve as fish bait, those of *Sonneratia*, for example, for the catfish *Arius maculatus*. Certain Amazon River fishes react positively to the audible 'explosions' of the ripe fruits of *Eperua rubiginosa*. Fossil evidence indicates that saurochory is very ancient. The giant Galapagos tortoise is important for the dispersal of local cacti and tomatoes. The name alligator apple for *Annona palustris* refers to its method of dispersal, an example of saurochory. Many birds and mammals, ranging in size from mice and kangaroo rats to elephants, eat and disperse seeds and fruits. In the tropics, chiropterochory (dispersal by large bats such as flying foxes, *Pteropus*) is particularly important. Fruits adapted to these animals are relatively large and drab in colour, with large seeds and a striking (often rank) odour.

Such fruits are accessible to bats because of the pagoda-like structure of the tree canopy, fruit placement on the main trunk, or suspension from long stalks that hang free of the foliage. Examples include mangoes, guavas, breadfruit, carob, and several fig species. In South Africa, a desert melon (*Cucumis humifructus*) participates in a symbiotic relationship with aardvarks—the animals eat the fruit for its water content and bury their own dung, which contains the seeds, near their burrows. Furry terrestrial mammals are the agents most frequently involved in epizoochory, the inadvertent carrying by animals of dispersal units. Burrlike seeds and fruits, or those diaspores provided with spines, hooks, claws, bristles, barbs, grapples, and prickles, are genuine hitchhikers, clinging tenaciously to their carriers.

Their functional shape is achieved in various ways—in cleavers, or goose grass (*Galium aparine*), and enchanter's nightshade (*Circaea lutetiana*), the hooks are part of the fruit itself; in common agrimony (*Agrimonia eupatoria*), the fruit is covered by a persistent calyx (the sepals, parts of the flower, which remain attached beyond the usual period) equipped with hooks; in wood avens (*Geum urbanum*), the persistent styles have hooked tips. Other examples are bur marigolds, or beggar's-ticks (*Bidens* species); buffalo burr (*Solanum rostratum*); burdock (*Arctium*); *Acaena*; and many *Medicago* species.

The last-named, with dispersal units highly resistant to damage from hot water and certain chemicals (dyes), have achieved wide global distribution through the wool trade. A somewhat different principle is employed by the so-called trample burrs, said to lodge themselves between the hooves of large grazing mammals. Examples are mule grab (*Proboscidea*) and the African grapple plant (*Harpagophytum*). In water burrs, such as those of the water nut *Trapa*, the spines should probably be considered as anchoring devices.

DISPERSAL BY ANTS

Mediterranean and North American harvester ants (*Messor, Atta, Tetramorium*, and *Pheidole*) are essentially destructive, storing and fermenting many seeds and eating them completely. Other ants (*Lasius, Myrmica*, and *Formica* species) eat the fleshy, edible appendage (the fat body or elaiosome) of certain specialized seeds, which they disperse.

Most myrmecochorous plants (species of violet, primrose, hepatica, cyclamen, anemone, corydalis, *Trillium*, and bloodroot) belong to the herbaceous spring flora of northern forests. Tree poppy (*Dendromecon*), however, is found in the dry California chaparral; *Melica* and *Centaurea* species, in arid Mediterranean regions. The so-called ant epiphytes of the tropics (*i.e.*, species of *Hoya, Dischidia, Aeschynanthus*, and *Myrmecodia*—plants that live in 'ant gardens' on trees or offer the ants shelter in their own body cavities) constitute a special group of myrmecochores that provide oil in seed hairs.

The ancestral forms of these hairs must have served in wind dispersal. The primary ant attractant of myrmecochorous seeds is not necessarily oil; instead, an unsaturated, somewhat volatile fatty acid is suspected in some cases. The myrmecochorous plant as a whole may also have specific adaptations; for example, cyclamen brings fruits and seeds within reach of ants by conspicuous coiling (shortening) of the flower stalk as soon as flowering is over.

DISPERSAL BY BIRDS

Birds, being preening animals, rarely carry burrlike diaspores on their bodies. They do, however, transport the very sticky (viscid) fruits of *Pisonia*, a tropical tree of the four-o'clock family, to distant Pacific islands in this way. Small diaspores, such as those of sedges and certain grasses, may also be carried in the mud sticking to waterfowl and terrestrial birds.

Synzoochory, deliberate carrying of diaspores by animals, is practiced when birds carry diaspores in their beaks. The European mistle thrush, *Turdus viscivorus*, deposits the viscid seeds of mistletoe (*Viscum album*) on potential host plants when, after a meal of the berries, it whets its bill on branches or simply regurgitates the seeds. The North American (*Phoradendron*) and Australian mistletoes (*Ameyema*) are dispersed by various birds, and the comparable tropical species of the plant family Loranthaceae by flowerpeckers (of the bird family Dicaeidae), which have a highly specialized gizzard that allows seeds to pass through but retains insects. Plants may also profit from the forgetfulness and sloppy habits of certain nut-eating birds that cache part of their food but neglect to recover everything, or drop units on their way to the hiding place. Best known in this respect are the nutcrackers (*Nucifraga*), which feed largely on the 'nuts' of beech, oak, walnut, chestnut, and hazel; the jays (*Garrulus*), which hide hazelnuts and acorns; the nuthatches; and the California woodpecker (*Balanosphyra*), which may embed literally thousands of acorns, almonds, and pecan nuts in bark fissures or holes of trees. Secondarily, rodents may aid in dispersal by stealing the embedded diaspores and burying them. In Germany an average jay may transport about 4,600 acorns per season, over distances of up to 4 km (2.5 miles). Woodpeckers, nutcrackers, and squirrels are responsible for a similar dispersal of *Pinus cembra* in the Alps near the tree line.

Most ornithochores (plants with bird-dispersed seeds) have conspicuous diaspores attractive to such fruit-eating birds as thrushes, pigeons, barbets (members of the bird family Capitonidae), toucans, and hornbills (family Bucerotidae), all of which either excrete or regurgitate the hard part undamaged. Such diaspores have a fleshy, sweet, or oil-containing edible part; a striking colour (often red or orange); no pronounced smell; a protection against being eaten prematurely in the form of acids and tannins that are present only in the green fruit; a protection of the seed against digestion—bitterness, hardness, or the presence of poisonous compounds; permanent attachment; and, finally, absence of a hard outer cover. In contrast to bat-dispersed diaspores, they occupy no special position on the plant. Examples are rose hips, plums, dogwood fruits, barberry, red currant, mulberry, nutmeg fruits, figs, blackberries, and others. The natural and abundant occurrence of *Euonymus* (cardinal's hat), which is a largely tropical genus, in temperate Europe and Asia, can be understood only in connection with the activities of birds. Birds also contributed substantially to the repopulation with plants of the island Krakatoa after the catastrophic eruption of 1883. Birds have made *Lantana* (originally American) a pest in Indonesia and Australia; the same is true of wild plums (*Prunus serotina*) in parts of Europe, *Rubus* species in Brazil and New Zealand, and olives (*Olea europaea*) in Australia.

Mimicry—the protection-affording imitation of a dangerous or toxic species by an edible, harmless one—is shown in reverse by certain bird-dispersed 'coral seeds' such as those of many species in the genera *Abrus, Ormosia, Rhynchosia, Adenanthera,* and *Erythrina*. Hard and often shiny red or black and red, many such seeds deceptively suggest the presence of a fleshy red aril and thus invite the attention of hungry birds.

SEEDS & FRUITS DISPERSED BY WIND

Like an endless army of parachutists released from an airplane, seeds and fruits travel the wind currents and gentle breezes of the earth, possibly colonizing a distant mountain slope or fertile valley. Literally hundreds of species in many plant families have adopted this remarkable method of dispersal, including a variety of ubiquitous plants that we recognize as 'weeds.' The answer to why some weedy composites (such as the European dandelion) have worldwide distributions is truly 'blowing in the wind' (to quote from the Peter, Paul and Mary song). Some of the ingenious adaptations for this method of wind dispersal include seeds that resemble parachutes, helicopters and gliders. In fact, one species reportedly inspired the design of some early aircraft. An astronomer friend of Mr. Wolffia once observed a strange formation of flying objects through his telescope. He was focusing on a squadron of tiny parachute seeds high above his house. And the entire plant body of wolffia (the world's smallest flowering plant) may be transported by powerful cyclonic storms. In the southeastern United States there are records of wolffia plant bodies less than one millimetre long being carried by a tornado, and they have even been reported in the water of melted hailstones.

AN INTRODUCTION TO THE BOTANY OF SEEDS

Seeds provide the vital genetic link and dispersal agent between successive generations of plants. Angiosperm seeds are produced and packaged in botanical structures called fruits which develop from the 'female' pistils of flowers. Immature seeds (called ovules) each contain a minute, single-celled egg enclosed within a 7-celled embryo sac. The haploid (1n) egg is fertilized by a haploid (1n) sperm resulting in a diploid (2n) zygote that divides by mitosis into a minute, multicellular embryo within the developing seed. A second sperm unites with 2 haploid polar nuclei inside a binucleate cell called the endosperm mother cell which divides into a mass of nutritive tissue inside the seed. In most seeds the embryo is embedded in this endosperm tissue which provides sustenance to the embryo during germination. In exalbuminous seeds (found in many plants such as the legumes), the endosperm tissue is already absorbed by the time you examine a mature seed within the pod, and the 2 white fleshy halves in the seed are really the cotyledons (components of the embryo). The 2 sperm involved in the double fertilization process originated within the pollen tube that penetrated the embryo sac. The pollen grain (and pollen tube) come from the 'male' organs (called anthers) on the same plant or different parental plants in a remarkable process known as pollination. Pollination is also accomplished by the wind (or water), and it may also involve insects in some of nature's most fascinating relationships between a plant and an animal. This is especially true of the amazing fig trees and their symbiotic wasps.

One of the important functions of seeds and fruits is dispersal; a mechanism to establish the embryo-bearing seeds in a suitable place away from their parental plants.

There are 3 main mechanisms for seed and fruit dispersal:

1. Hitchhiking on animals,
2. Drifting in ocean or fresh water, and
3. Floating in the wind.

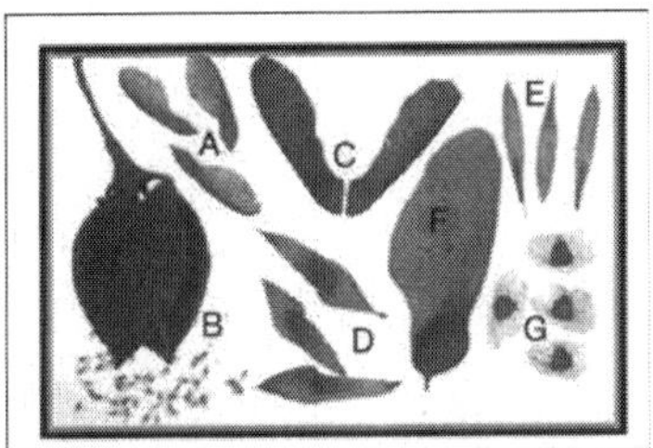

Wind-dispersed seeds & fruits in different plant families:

- *Helicopters:* A. Box Elder (Acer negundo, Aceraceae); C. Big-Leaf Maple (Acer macrophyllum, Aceraceae); E: Evergreen Ash (Fraxinus uhdei, Oleaceae); F. Tipu Tree (Tipuana tipu, Fabaceae).
- *Flutterer/Spinners:* B. Empress Tree (Paulownia tomentosa, Scrophulariaceae); D. Tree Of Heaven (Ailanthus altissima, Simaroubaceae); G. Jacaranda (Jacaranda mimosifolia, Bignoniaceae).

Gliders

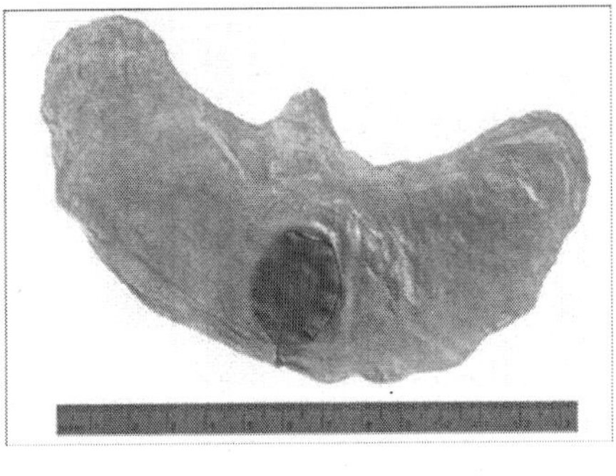

Fig. The Remarkable Winged Seed of the Tropical Asian Climbing Gourd Alsomitra Macrocarpa.

Gliders include seeds with 2 lateral wings that resemble the wings of an airplane. They become airborne when released from their fruit and sail through the air like a true glider. One of the best examples of this method is Alsomitra macrocarpa, a tropical vine in the Gourd Family (Cucurbitaceae) native to the Sunda Islands of the Malay Archipelago. Football-sized gourds hang from the vine high in the forest canopy, each packed with hundreds of winged seeds. The seeds have two papery, membranous wings, with combined wingspans of up to 5 inches (13 cm). They reportedly inspired the wing design of some early aircraft, gliders and kites. Although the seeds vary in shape, some of the most symmetrical ones superficially resemble the shape of the 'flying wing' aircraft or a modern Stealth Bomber. According to Peter Loewer, the aerodynamic seeds spiral downward in 20 foot (6 meter) circles, although a gust of wind would probably carry them much farther away.

Helicopters (Whirlybirds)

Helicopters (also called Whirlybirds) include seeds or one-seeded fruits (samaras) with a rigid or membranous wing at one end. The wing typically has a slight pitch (like a propeller or fan blade), causing the seed to spin as it falls. Depending on the wind velocity and distance above the ground, helicopter seeds can be carried considerable distances away from the parent plant. The spinning action is similar to auto-rotation in helicopters, when a helicopter 'slowly' descends after a power loss.

Numerous species of flowering trees and shrubs in many diverse and unrelated plant families have evolved this ingenious method of seed dispersal, good examples of convergent evolution. Representative examples of helicopter seeds and one-seeded fruits (called samaras) include the Maple Family (Aceraceae): Maples and box elder (Acer); Olive Family (Oleaceae): Ash (Fraxinus); Legume Family (Fabaceae): Tipu tree (Tipuana tipu); and the Protea Family (Proteaceae): Banksia and Hakea.

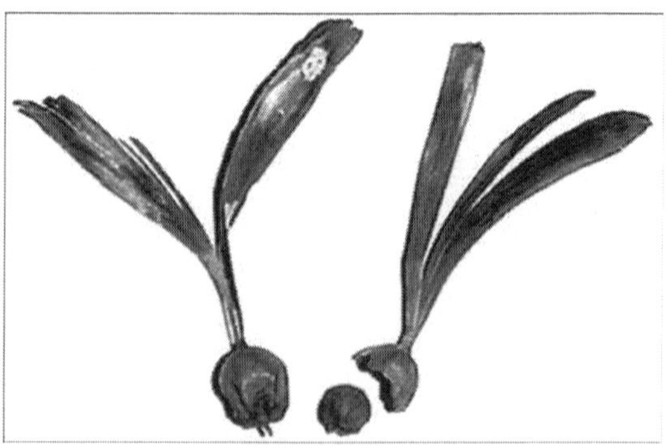

Fig. An Interesting One-seeded Winged Fruit that Spins as it Falls through the Air. It is called 'Whirling Nut' and belongs to the Genus Gyrocarpus in the Family Gyrocarpaceae.

Note: This genus of tropical trees, shrubs and lianas is often placed in the hernandia family (Hernandiaceae). The unusual fruit shown above was collected and photographed at Ho'omaluhia Botanical Garden on the windward side of Oahu in the Hawaiian Islands. Identification provided by Ricarda Riina, Botany Department, University of Wisconsin.

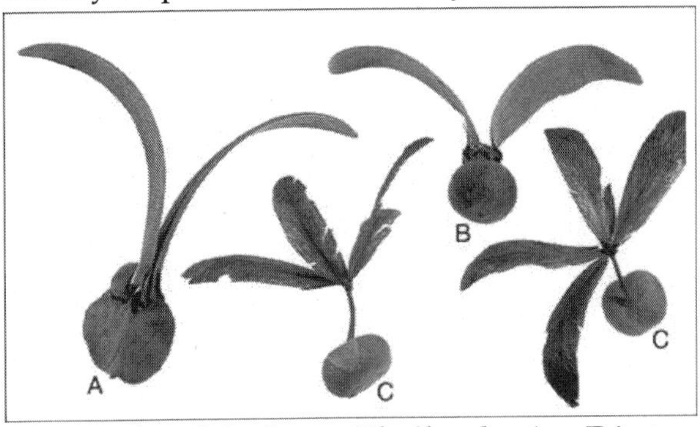

Spinning fruits (seeds) from Thailand: A. Diptocarpus alatus (Diptocarpaceae). B. Diptocarpus obtusifolius. C. Gluta (Melanorrhoea) usitata (Anacardiaceae). The latter species is also called the Burmese lacquer tree and is well-known as a source of lacquer. The resin is chemically similar to the Japanese lacquer tree. Like the Japanese lacquer tree and poison oak, the resin canals also contain urushiol, a mixture of toxic phenolic compounds that cause a cell-mediated immune response in some people. Identifications courtesy of Dr. Tomiki Sando, Thailand.

Although they are classified as gymnosperms with naked seeds arising from woody cones rather than flowers, the Pine Family (Pinaceae) contains many genera with winged seeds, including Pinus (Pine), Abies (fir), Picea (spruce), Tsuga (hemlock), and many additional genera. When shed from cones high on upper branches, they fly over slopes and across deep canyons. The natural reforestation of conifers following fire is proof of the flying ability of seeds from nearby forested slopes.

Maples have a double or twin samara composed of 2 winged one-seeded fruits (double samara) joined together at their bases. When they break apart, each winged fruit flies like a typical helicopter seed. Although the Legume Family (Fabaceae) is the third largest plant family with over 18,000 described species, the vast majority of legumes do not have winged seeds or fruits. The South American tipu tree (Tipuana tipu) is a notable exception, with beautiful yellow blossoms that give rise to pendant, samara-like legumes, each with a large wing on the lower end. The dried, winged legumes spin so neatly in the air that they could be marketed as a child's toy.

The remarkable Protea Family (Proteaceae) of Australia contains some truly amazing genera with winged seeds, including Banksia and Hakea. Although they are flowering plants, banksias produce a dense flower cluster (inflorescence) that gives rise to a cone-like structure containing many woody carpels. Each carpel bears 2 winged seeds and the entire cone-like structure superficially resembles a pine cone. In fact, some banksias release their seeds following fire and even resprout from subterranean lignotubers like chaparral shrubs.

Flutterer/Spinners

The native range of hopseed bush (Dodonea viscosa), a member of the Soapberry Family (Sapindaceae), extends from Arizona to South America. It is also commonly cultivated in southern California. The papery, winged fruits flutter and spin in the air, and may be carried short distances by the wind.

The jacaranda tree (Jacaranda mimosifolia) of northwestern Argentina. Like many other members of the Bignonia Family (Bignoniaceae), the papery, winged seeds flutter and spin as they are carried by the wind.

Although their mode of dispersal is similar to single-winged helicopter seeds, the flutterer/spinners include seeds with a papery wing around the entire seed or at each end. When released from their seed capsules they flutter or spin through the air. Whether they spin or merely flutter depends on the size, shape and pitch of the wings, and the wind velocity.

This method of wind dispersal is found in numerous species of flowering plants in many different plant families. Some examples of flutterer/spinner seeds include the Quassia Family (Simaroubaceae): Tree of heaven (Ailanthus altissima); Figwort Family (Scrophulariaceae): Empress tree (Paulownia tomentosa); Bignonia Family (Bignoniaceae): Jacaranda (Jacaranda mimosifolia), catalpa (Catalpa speciosa), desert willow (Chilopsis linearis), yellow bells (Tecoma stans), bower vine (Pandorea jasminoides), violet trumpet vine (Clytostoma callistegioides), and the fabulous trumpet trees (Tabebuia serratifolia and T. ipe); Elm Family (Ulmaceae): American and Chinese elms (Ulmus americana and U. parvifolia); Soapberry Family (Sapindaceae): Hop seed (Dodonea viscosa); and the Goosefoot Family (Chenopodiaceae): Four-wing saltbush (Atriplex canescens).

Any discussion of flutterer/spinners would not be complete without mentioning the quipo tree (Cavanillesia platanifolia), a massive rain forest tree in the bombax family (Bombacaeae) native to Panama. The enormous winged fruits of the quipo tree flutter through the air, carpeting the ground beneath the huge canopy of this striking tropical tree.

Fig. The quipo tree (Cavanillesia platanifolia), a remarkable rain forest tree in the bombax family (Bombacaceae) with huge winged fruits. This massive tree is native to Panama.

Some of the most beautiful flowering trees of the New World tropics belong to the Bignonia Family (Bignoniaceae). They typically produce long, slender (cigar-shaped) seed capsules containing masses of flat seeds with papery wings at each end. [The beautiful jacaranda of Argentina has flattened, circular seed capsules.] The lovely yellow bells (Tecoma stans) is native to Mexico and the Caribbean region, and is the official flower of the U.S.

Virgin Islands. Some of the South American trumpet trees, including the pink-flowered Tabebuia avellanedae (listed as T. ipe in some references) and the yellow-flowered Tabebuia serratifolia, are also called ironwoods or axe-breakers (quebrachos) because of their dense, hard wood. The latter species is called 'pau d'arco' and its wood actually sinks in water, with a specific gravity of 1.20.

In South America, trumpet trees drop their leaves during the dry season and produce a profusion of pink or yellow blossoms. The crowns of these huge timber trees resemble gigantic floral bouquets in the midst of the forest. As with so many tropical species, some of the trumpet trees inhabit rain forest areas that are seriously threatened by slash and burn agriculture, large plantations of exportable products, and the general annihilation of the South American rain forests.

Other South American species of Tabebuia are also referred to as pau d'arco, including the pink-flowered T. impetiginosa and T. avellanedae. According to The New York Botanical Garden Encyclopedia of Horticulture Volume 10, 1982, T. avellanedae is a synonym for T. impetiginosa, and T. ipe ' is so closely similar to T. impetiginosa that it can scarcely be more than a variety of that species.' These attractive pink-flowered species are commonly used as landscape trees in temperate regions.

The powdered inner bark of these pink-flowered species of pau d'arco is sold as a popular herbal remedy that reportedly stimulates the immune system. According to a book by Kenneth Jones, this valuable herb has been proven successful in the treatment of certain cancers, allergies associated with the Candida yeast syndrome, and in disorders involving a weakened immune system.

Parachutes

Parachutes include seeds or achenes (one-seeded fruits) with an elevated, umbrella-like crown of intricately-branched hairs at the top, often produced in globose heads or puff-like clusters.

Fig. An Individual Parachute of Western Salsify (Tragopogon dubius) Showing an Umbrella-like, Plumose Crown of Hairs (pappus) above a Slender one-seeded Fruit (called an Achene). These Fragile Units can become Airborne with the Slightest Gust of Wind, and can Literally Sail Across Valleys and Over Mountain Slopes.

Fig. Western Salsify or Goatsbeard (Tragopogon Dubius) Showing Dense, Puff-like Cluster of Numerous Parachute Seeds (One-seeded Achenes). Each Achene has an Umbrella-like Crown of Plumose Hairs and may Literally be Carried into the Atmosphere by Strong Ascending Air Currents.

The slightest gust of wind catches the elaborate crown of plumose hairs, raising and propelling the seed into the air like a parachute. This is the classic mechanism of dispersal for the Eurasian dandelion (Taraxacum officinale) and includes numerous weedy and native members of the Sunflower Family (Asteraceae).

A giant Eurasian version of the dandelion called salsify or goat's beard (Tragopogon dubius), is one of the most successful wind-travellers in North America. Its seeds have literally blown across mountain ranges, colonizing vast fields of open land in the western United States. Three weedy species of salsify (T. dubius, T. pratensis and T. porrifolius) have been introduced into the western United States, 2 with yellow dandelion-type flowers and one with purple flowers. The latter, purple-flowered species (T. porrifolius) has a large, edible tap root with a flavour resembling oysters, hence the name 'oyster plant.'

In some parachutes, the crown of silky hairs arises directly from the top of the seed (not on an umbrella-like stalk). Again, the Sunflower Family (world's largest plant family with about 24,000 described species) contains many weedy representatives with this type of parachute seed. One of the most troublesome weeds of farm land in the western United States is wild or thistle artichoke (Cynara cardunculus). The large seed head of this weedy composite releases hundreds of parachute seeds which fly through the air and invade vast areas of grazing land with spiny, perennial bushes that literally take over.

The large leaf stalks (resembling giant celery stalks) are edible and are sold under the name of 'cardoon.' Populations of wild artichoke often contain so much variation between spiny and non-spiny plants, that some experts believe that they belong to one variable species. In fact, some botanists believe that the cultivated artichoke (C. scolymus) may be a cultivated variety of the wild C. cardunculus. Incidentally, the delicious artichoke is really a cooked flower head in which the outer bracts (phyllaries) and central basal portion (receptacle) are dipped in butter and eaten.

Fig. Brown Puffs (Stebbinoseris heterocarpa), formerly Microseris Heterocarpa, an Interesting Member of the Sunflower Family. A Closely Related Species is called Silver Puffs (Uropappus lindleyi syn. Microseris lindleyi). In Stebbinoseris the Pappus Paleae are Bifid at the Apex. Unlike the Weedy Dandelions (Taraxacum) and Salsify (Tragopogon), this is a Native Species in California.

Another plant family which has evolved this parachute method of seed dispersal is the Milkweed Family (Asclepiadaceae). Hundreds of parachute seeds (each with a tuft of silky hairs) are produced within large, inflated pods called follicles. So abundant are the silky hairs, that they were actually collected and used as a substitute for kapok during World War II. Kapok comes from masses of silky hairs that line the seed capsules of the kapok tree (Ceiba pentandra), an enormous rain forest tree of Central and South America.

Kapok is used primarily as a waterproof filler for mattresses, pillows, upholstery, softballs, and especially for life preservers. The floss silk tree (Chorisia speciosa), another member of the Bombax Family (Bombaceae) also produces large seed capsules lined with masses of silky hairs. This tree with its distinctive thorny trunk and showy pink flowers is commonly planted in southern California. The seeds of kapok and floss silk trees are embedded in these silky masses which aid in their dispersal by wind.

The Dogbane Family (Apocynaceae) also includes members with seed pods (follicles) and parachute seeds similar to those of milkweeds. One of the best examples is Nerium oleander, a drought-resistant, Mediterranean shrub planted throughout southern California. The foliage contains a powerful cardiac glycoside that can permanently relax the heart muscle.

Fig. Parachute Seeds Escaping from the Follicle of Nerium Oleander. The Crown of Silky Hairs Arises Directly from the Top of the Seed (not on an Umbrella-like Stalk. Unlike the Seed-bearing Achenes of the Sunflower Family (Asteraceae), these are True Seeds.

SPECIFIC GRAVITY

Probably the best way to appreciate the relative hardness of different woods is the concept of 'specific gravity,' a numerical scale based on 1.0 for pure water. Without getting too mathematical, the specific gravity of a substance can easily be calculated by dividing its density (in grams per cubic centimeter) by the density of pure water (one gram per cubic centimeter).

The brilliant Greek mathematician and inventor Archimedes discovered over 2,100 years ago that a body in water is buoyed up by a force equal to weight of the water displaced. Archimedes reportedly came upon this discovery in his bathtub, and ran out into the street without his clothing shouting 'Eureka, I have found it.' Since one gram of pure water occupies a volume of one cubic centimetre, anything having a specific gravity greater than 1.0 will sink in pure water.

The principles of buoyancy and specific gravity are utilized in many ways, from scuba diving and chemistry to the hardness of dry, seasoned wood. Some of the heaviest hardwood trees and shrubs of the United States have specific gravities between 0.80 and 0.95; including shagbark hickory (Carya ovata), persimmon (Diospyros virginiana) and ironwood (Ostrya virginiana) of the eastern states, and canyon live oak (Quercus chrysolepis), Engelmann oak (Q. engelmannii), hollyleaf cherry (Prunus ilicifolia) and Santa Cruz Island ironwood (Lyonothamnus floribundus ssp. asplenifolius) of southern

California. Although some of these trees are called ironwoods, their dense, dry wood will still float in water. Since the pure cell wall material (lignin and cellulose)) of wood has a density of about 1.5 grams per cubic centimetre, even the world's heaviest hardwoods generally have specific gravities less than 1.5 due to tiny pores (lumens) within the cell walls. True ironwoods include trees and shrubs with dry, seasoned woods that actually sink in water, with specific gravities greater than 1.0. They include lignum vitae (Guaicum officinale, 1.37); quebracho (Schinopsis balansae, 1.28); pau d'arco (Tabebuia serratifolia, 1.20); knob-thorn (Acacia pallens, 1.19); desert ironwood (Olneya tesota, 1.15); and ebony (Diospyros ebenum, 1.12). To appreciate the weight of these hardwoods, compare them with tropical American balsa (Ochroma pyramidale), one of the softest and lightest woods with a specific gravity of only 0.17.

Cottony Seeds & Fruits

Cottony seeds and fruits include seeds and minute seed capsules with a tuft (coma) of cottony hairs at one end, or seeds embedded in a cottony mass. Some of the examples in this group are very similar in function to parachute seeds, but probably are not carried as far by the wind. Many plant families have this type of wind dispersal, including the Willow Family (Salicaceae): Willows (Salix) and Cottonwoods (Populus); Cattail Family (Typhaceae): Cattails (Typha); Evening Primrose Family (Onagraceae): Willow-Herb (Epilobium) and California fuchsia (Zauschneria); Bombax Family (Bombaceae): Kapok tree (Ceiba pentandra) and floss silk tree (Chorisia speciosa); and the Sycamore Family (Platanaceae): Sycamore (Platanus).

In the California sycamore (Platanus racemosa), a common riparian (streamside) tree throughout the state, the one-seeded fruits (achenes or nutlets) are produced in dense, globose heads. The spherical heads hang from branches like little balls. Individual achenes have a tuft of hairs at the base which probably helps in their wind dispersal. Seeds of the South American kapok tree (Ceiba pentandra) and floss silk tree (Chorisia speciosa) are embedded in dense masses of silky hairs inside large woody capsules. This undoubtedly helps to disperse the seeds when seed-bearing masses of hair are carried by the wind. In tropical regions of the New World, the kapok grows into an enormous rain forest tree with a massive buttressed trunk. Kapok hairs are coated with a highly water-resistant, waxy cutin layer. The empty lumen (cavity) inside each hair is larger the cotton hairs; hence, the hairs are lighter. Unlike cotton hairs, kapok is difficult to spin and is not made into textiles. It is used primarily as a waterproof filler for mattresses, pillows, upholstery, softballs, and especially for life preservers. A kapok-filled life jacket can support 30 times its own weight in water.

One fuzzy brown cattail spike may contain a million tiny seeds. Each seed has a tuft of silky white hairs and is small enough to pass through the 'eye' of an ordinary sewing needle. They are shed in clouds of white fluff and float through the air like miniature parachutes. A cattail marsh covering one acre may produce a trillion seeds, more than 200 times the number of people in the world. The fluffy seeds have been used for waterproof insulation and the buoyant filling of life jackets. In addition, each plant produces billions of wind-borne pollen grains; in fact, so much pollen that it was used as flour by North American Indians and made into bread. Cottonwoods and willows also produce masses of seeds, each with a tuft of soft, white hairs. Since they are dioecious, with pollen-bearing male and seed-bearing female trees in the population, only female trees produce the actual cotton. During late spring and summer in the western United States, the cottony fluff from cottonwoods resembles newly fallen snow. Because the wind-blown fluff can be quite messy in cultivated parks and gardens, male trees are generally planted. The discriminatory label of 'cottonless cottonwood' refers to a male tree.

Miscellaneous

Squirrel-Tail Grass (Elymus elymoides), formerly named Sitanion hystrix is an attractive grass native to the mountains and plains of the western United States. Seed-bearing sections (spikelets) of the flower spike (containing one-seeded fruits called grains and very long awns) are carried short distances by the wind. Although not as efficient fliers, the long awns function like the parachute bristles (pappus) of composites.

This miscellaneous category of wind-blown seeds and fruits includes plants that really don't fit the above 5 categories. The Grass Family (Poaceae) includes a number of species with plumose flower stalks that fragment into seed-bearing spikelets that blow into the wind. Some of these species have become troublesome weeds in southern California, including the South African fountain grass (Pennisetum setaceum). Although this tufted perennial makes an attractive, drought-resistant landscaping plant along walkways and roads, it is becoming a widespread weed in disturbed areas of San Diego County. Another species, called squirrel-tail grass (Elymus elymoides), resembles a weedy introduced grass, but it is actually a native perennial of dry, rocky mountains and open land in the western United States. To appreciate its airborne seeds, you really must see this grass during a strong gust of wind on the eastern slopes of the Sierra Nevada during late summer. Mountain mahogany (Cercocarpus minutiflorus), a native shrub in the chaparral of southern California, produces a rather unique wind-blown fruit. The one-seeded fruit (achene) has a persistent, feathery style that glistens in the sunlight. Although they usually don't travel very far, the achenes are blown into the air by strong gusts of wind during the dry, fire season of late summer and fall. This species is not related to the West Indian mahogany (Swietenia mahagoni) or the Honduran mahogany (S. macrophylla), members of the true Mahogany Family (Meliaceae). Mountain mahogany actually belongs to the Rose Family (Rosaceae) and produces very hard wood that sinks in water when dry. In fact, the wood of a montane species (C. ledifolius), has a specific gravity of 1.12, as heavy and dense as ebony (Diospyros ebenum).

Tumbleweed (Russian Thistle)

The common tumbleweed or Russian thistle is a rounded, bushy annual introduced into the western United States from the plains of southeastern Russia and western Siberia in the late 1800s. The name 'thistle' comes from the stiff, sharp-pointed, awl-shaped leaves. Although it is depicted in songs of the old west, this species is a naturalized weed in North America. It is listed in most older references as Salsola kali or S. pestifer; however, the Jepson Flora of California (1993) lists it as S. tragus. Russian thistle belongs to the goosefoot family (Chenopodiaceae), along with many weedy species and some valuable vegetables, including beets (Beta vulgaris), goosefoot (Chenopodium album) and spinach (Spinacia oleracea).

Tumbleweed is a prolific seeder and rapid seed germination and seedling establishment occurs after only a brief and limited rainy season. A single plant may produce 20,000 to 50,000 seeds within numerous small fruits, each surrounded by a circular, papery border. Mature plants readily break off at the ground level and are pushed along by strong gusts of wind. As they roll along hillsides and valleys, the seeds are scartered across the landscape. Tumbleweeds often pile up in wind rows along fences and buildings. This is a troublesome weed in agricultural areas because it literally covers the farm land with bushy, prickly shrubs. One interesting use for this plant in arid regions of the American southwest is for a 'snowman' at Christmas time. Three proportionally sized tumbleweeds are used to make the head, thorax and main body of a 'snowman.' Another suggested use is to compress tumbleweeds into logs and use them for firewood.

DISPERSAL BY WATER

Many marine, beach, pond, and swamp plants have waterborne seeds, which are buoyant by being enclosed in corky fruits or air-containing fruits or both; examples of these plants include water plantain, yellow flag, sea kale, sea rocket, sea beet, and all species of Rhizophoraceae, a family of mangrove plants. Sea dispersal of the coconut palm has been well proved; the fibrous mesocarp of the fruit, a giant drupe, provides buoyancy. Once the nuts are ashore, the mesocarp also aids in the above ground germination process by collecting rainwater; in addition, the endosperm has in its 'milk' a provision for seedling establishment on beaches without much fresh water. A sea rocket species with seeds highly resistant to seawater is gaining a foothold on volcanic Surtsey Island, south of Iceland. Purple loosestrife, monkey flower, *Aster tripolium,* and *Juncus* species (rushes) are often transported by water in the seedling stage. Rainwash down mountain slopes may be important in tropical forests. A 'splashcup mechanism,' common in fungi for spore dispersal, is suggested by the open fruit capsule with exposed small seeds in the pearlwort (*Sagina*) and mitrewort (*Mitella*). Hygrochasy, the opening of fruits in moist weather, is displayed by species of *Mesembryanthemum, Sedum,* and other plants of dry environments.

8

Seed Management and Dynamics System

According to ENHRUM data, between 1997 and 2001, 0.5 per cent of Mexican rural farmers sowed maize seed brought from the US, but none of them conserved this seed in 2002. Nearly 3 per cent of farmers sowed maize grain obtained in Diconsa, the public retail network, at least once during the same 5-year period, but only 0.5 per cent of seed lots sown in 2002 came from this source. Seed obtained from government agencies was nearly as common as Diconsa's, while the formal seed system and other sources of grain each account for 10 times more seed. Seed exchange with other farmers through informal seed systems was overwhelmingly the main source of seed across Mexico. Its importance is much greater in the southeast than in the north, where the seed industry and other institutional sources are also significant.

SEED SYSTEM

Success in seed production is judged not only by the quantity of good seed, but more importantly, by the process through which seed is produced, and the capacity to sell this seed and make it accessible to intended end-users on a regular basis. This is where economics becomes relevant in attempting to seek answers to questions relating to cost-efficiency in seed production, distribution and marketing. The seed industry, with its affiliated institutions and services together,constitute the seed system or seed sector. In the organized seed sector, linkages within the production—distribution–use chain and external institutions are more structured. There are several dimensions to seed system analysis which broadly cover technical, institutional, financial, economic, social, environment, policy, and commercial issues. These are not separated in function, but are rather inter-dependent.

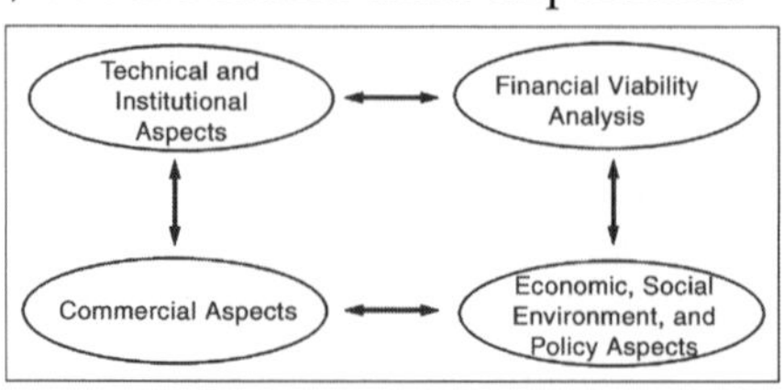

Fig. Complementary Components of the Seed System

Since little is written on the subject of economics in seed production, this book has been prepared to meet two specific purposes: first, to provide information on the economic basis of seed operations and second, to suggest guiding principles in running cost-effective seed businesses. The text is intended primarily for use by technical staff and managers in seed programs. Information provided is of relevance to developing countries at different stages of seed industry growth. Therefore, a choice has to be made between economic principles, examples, and alternatives given to suit specific situations.

The researchers presented findings from the project and provided contrasting lessons from their respective studies, which will be produced as FAC Working Papers and a special issue of the IDS Bulletin on the 'Politics of Seed' (July 2011). The proceedings of the conference will also be published by EIAR, with support from FAC.

A wide ranging open plenary discussion closed out the second day of the conference. This touched on a number of points raised by the FAC STI presenters, including the politics behind seed R&D strategies and policies that either open up or close down alternative innovation pathways, the potential and limitations of the agro-dealer model, the challenges of establishing a private sector seed sector with a strong role for farmer-led 'local seed businesses'.

FORMAL SEED SYSTEMS

Seed is one of the critical inputs in groundnut production. Diffusion of improved seeds to farmers is an efficient way to increase production of the crop in Nigeria. Although both the formal and informal seed Systems operate side-by-side, the informal seed sector is much more important, supplying more than 75 per cent of the seed requirements for groundnut in the country. This situation which is reflective of low economic activity and lack of professionalism among the operators of the seed sector can be improved if there is a complementarity between the formal and informal seed sectors. This will encourage capacity building in the development of technical and entrepreneurial skills, and management strategies among stakeholders in groundnut seed production. Support from government agencies, seed companies and non-governmental organizations will also create awareness about availability and potentials of improved varieties. The integration of formal and informal seed Systems in Nigeria can only guarantee improved seed supply to the groundnut farmer.

TECHNICALITY IN PRODUCTION

REQUIREMENT OF SEED PRODUCTION

The basic requirement of seed production is availability of improved varieties/hybrids and their demand among the farmers. Preferably the variety should be realsed and notified for certified seed production, however, any kind/variety can be multiplied for Truthfully Labeled Seed (TFL).

ISOLATION REQUIREMENTS

The cucurbits are cross pollinated in nature and honeybees are major pollinator, thus for pure seed production an isolation distance all around seed field is necessaryto separate it from fields of other varieties, fields of the same variety not confirming to varietal purity requirement. The isolation distance of 400 m for C.S. and 800 m for F.S. and at least1000 m isolation is required for breeder seed production.

It is important to mention that muskmelon, longmelon and snapemelon (phoot) can cross with each other. Similarly, species of genus Cucurbita have the risk of crossability among them. The cucumber can cross easily with its wild relative Cucumis hardwickee found in wild form in sub-mountanious regions of Himalayas.

CHOICE OF SEASON AND AREAS OF SEED PRODUCTION

Seed crop should be raised in such a seasons which remain dry at the time of seed maturity and seed extraction. Rainy season is preferred over summer season for raising seed crop.

Locations are also important in seed production with reference to seed yield and quality of seed. To harness the advantage of climate, private sector seed companies are organizing their seed production in these areas.

MAJOR CUCURBITS SEED PRODUCTION REGIONS IN INDIA

Crop	Areas	Region
Muskmelon, Longmelon	Punjab & Haryana	Northern Region
Bittergourd	Eastern U.P., Faizabad and Jaunpur	Northern Region
Cucumber and Muskmelon	Azamgarh, Ballia and Gonda in U.P. Jalana (Maharashtra)	Northern Region Central Region
Pumpkin and Ridge gourd	West Bengal (South West)	Eastern Region
Sponge gourd, Watermelon, Cucumber	Telingana, Karool and Vijayawada (North A.P.) Ranibenur (Karnataka)	Southern Region Southern Region
Watermelon, Bottle gourd and Bitter gourd	Costal districts of A.P.	Southern Region

Roguing

Seed crop is to be monitored at various stages of crop growth for removal of off-type and obviously should be carried out before flowering to avoid natural cross-pollination. However, fruit set and complete fruit development stages are also important. The crop-wise main features described here under for effective roguing.

- *Muskmelon:* Fruit shape, colour, rind colour, skin (netted/plain), flesh colour (orange/red), TSS and cavity size.
- *Watermelon*: Fruit shape, colour, rind colour, flescolou(red/yellow/white).
- *Longmelon*: Fruit shape, colour, bitterness.
- *Cucumber*: Fruit shape, colour, presence of spines, spines colour, colour of ripen fruit (green, yellow, white or orange)
- *Pumpkin (Squash, summer)*: Fruit shape, colour, flesh colour.
- *Gourds (Bottle gourd, Bitter gourd and Luffa etc.)*: Fruit shape, colour, stripe, neck etc.

MATURITY OF FRUIT

Cucurbits takes fairly long time to attain harvestable maturity. The maximum period is required in crops like pumpkin (Cucurbita moschata), ashgourd and watermelon, however, muskmelon, round melon and bitter gourd take relatively lesstime.

The maturity also influenced by the environmental factors and crop management (trailing etc.). The maturity period is shorter in summer season than rainy season.

HARVESTABLE MATURITY IN CUCURBITS IN INDIA

Besides days to maturity, some other parameters like change in colour also used as crietaria of maturity index.

- *Cucumber and summer squash*: Fruit turn pale yellow to golden yellow and attached with plant.
- *Pumpkin*: Fruit reddena and seeds inside the shell breake readily from pulp.
- *Muskmelon*: Full slip stage.
- *Watermelon*: Fruits are ready for harvest when they reach edible maturity, fruit colour change from green/white to pale yellow of under side of the fruit.
- *Bitter gourd and snake gourd*: Fruit turn to bright yellow.
- *Bottle gourd*: At maturity fruit colour fade to starw freen or pale yellow.
- *Luffa*: Complete drying/fruit turn to gray colour.

Crop	Variety	Period in Days (Seed to Seed)
Bitter gourd	Priya, Pusa Vishesh	65 days
Snake gourd	TA-19	65 days
Bottle gourd	Arka Bahar,	75 days
	Pusa Naveen	80 days
Pumpkin	Ambali	90 days
	Pusa	(Rainy season)
	Vishwas/Vikas	100 days
Ash gourd	Local	90 days
Cucumber	Mudikode Local	70 days
Watermelon	Sugar Baby	85 days
Muskmelon	Pusa Madhuras	80 days

SEED EXTRACTION

There are two method of seed extraction employed in cucurbits:

1. *Dry method*: The dried fruits are cut from one side and the seeds comes out from the fruit *e.g.*, sponge gourd, ridge gourd, snake gourd.
2. *Wet method*: This method is employed for seeds extraction of cucumber, muskmelon, watermelon, ash gourd, bitter gourd, round melon and long melon. The fruit of cucumber and bitter gourd, summer squash and long melon are cut longitudianlly and seed is scooped out while fruit of muskmelon and pumpkin are cut into two piece and seed is scooped out from cavity. However, in case of watermelon and ash gourd whole central portion are manually scooped out and macerated to separate the seed from pulp. In wet method, the seed extraction done by three ways:
   - *Mechanical Extraction:* In this method the fruits are cut into pieces and macerated by machine. The seeds are separated out from pulp by floating with water. This method is quick, less expensive and seeds retain good lusture, but require good amount of water. This method is applicable in bottlgourd, watermelon, roundmelon and ash gourd.
   - *Natural Fermentation*: The scooped material kept in wooden/ plastic or steel vesel for 48 hours at room temperature and stired 2-3 times and then seed is washed throughly with water 2-3 times. The main problem with this method are discolouration and poor lusture of seed.
   - *Chemical Extraction*: 25-30 ml. of HCL or 8-10 ml. of commercial H_2SO_4 added per 5 kg of pulp and some quantity of water is

mixed, stiring of pulp is done to enhance to separation and left for 30 minutes. The impurities will float and seed will sink. The seed should be washed throughly with clean water. This is quick method but accuracy of acid and time is important.

CONQUEST OF PRODUCTION

Even today we see men and women denying themselves necessaries to acquire mere trifles, to obtain some particular gratification, or some intellectual or material enjoyment. A Christian or an ascetic may disapprove of these desires for luxury; but it is precisely these trifles that break the monotony of existence and make it agreeable. Would life, with all its inevitable sorrows, be worth living, if besides daily work man could never obtain a single pleasure according to his individual tastes?

If we wish for a Social Revolution, it is no doubt in the first place to give bread to all; to transform this execrable society, in which we can every day see robust workmen dangling their arms for want of an employer who will exploit them; women and children wandering shelterless at night; whole families reduced to dry bread; men, women, and children dying for want of care and even for want of food. It is to put an end to these iniquities that we rebel.

But we expect more from the Revolution. We see that the worker compelled to struggle painfully for bare existence, is reduced to ignorance of these higher delights, the highest within man's reach, of science, and especially of scientific discovery; of art, and especially of artistic creation. It is in order to obtain these joys for all, which are now reserved to a few; in order to give leisure and the possibility of developing intellectual capacities, that the social revolution must guarantee daily bread to all. After bread has been secured, leisure is the supreme aim.

No doubt, nowadays, when hundreds and thousands of human beings are in need of bread, coal, clothing, and shelter, luxury is a crime; to satisfy it the worker's child must go without bread! But in a society in which all can eat sufficiently the needs which we consider luxuries to- day will be the more keenly felt. And as all men do not and cannot resemble one another (the variety of tastes and needs is the chief guarantee of human progress) there will always be, and it is desirable that there should always be, men and women whose desire will go beyond those of ordinary individuals in some particular direction.

Everybody does not need a telescope, because, even if learning were general, there are people who prefer examining things through a microscope to studying the starry heavens. Some like statues, some pictures. A particular individual has no other ambition than to possess an excellent piano, while another is pleased with an accordion. The tastes vary, but the artistic needs exist in all. In our present, poor capitalistic society, the man who has artistic needs cannot satisfy them unless he is heir to a large fortune, or by dint of hard work appropriates to himself an intellectual capital which will enable him to take up a liberal profession. Still he cherishes the *hope* of some day satisfying his tastes more of less, and for this reason he reproaches the idealist Communist societies with having the material life of each individual as their sole aim.— "In your communal stores you may perhaps have bread for all," he says to us, "but you will not have beautiful pictures, optical instruments, luxurious furniture, artistic jewelry—in short, the many things that minister to the infinite variety of human tastes. And in this way you suppress the possibility of obtaining anything besides the bread and meat which the commune can offer to all, and the grey linen in which all your lady citizens will be dressed". These are the objections which all communist systems have to consider, and which the founders of new societies, established in American deserts, never understood. They believed that if the community could procure sufficient cloth to dress all its members, a music hall in which the "brothers" could strum a piece of music, or act a play from time to time, it was enough. They forgot that the feeling for art existed in the agriculturist as well as in the burgher, and, notwithstanding that the expression of artistic feeling varies according to the difference in culture, in the main it remains the same. In vain did the community guarantee the common necessaries of life, in vain did it suppress all education that would tend to develop individuality, in vain did it eliminate all reading save the Bible. Individual tastes broke forth, and caused general discontent; quarrels arose when somebody proposed to buy a piano or scientific instruments; and the elements of progress flagged. The society could only exist on condition that it crushed all individual feeling, all artistic tendency, and all development. Will the anarchist Commune be impelled by the same direction? Evidently not, if it understands that while it produces all that is necessary to material life, it must also strive to satisfy all manifestations of the human mind.

SEED YIELD

ROLE OF SOCIETY

Is it a dream to concieve a society in which—all having become producers, all having received an education that enables them to cultivate science or art, and all having leisure to do so—men would combine to publish the works of their choice, by contributing each his share of manual work? We have already hundreds of learned, literary, and other societies; and these societies are nothing but voluntary groups of men, interested in certain branches of learning, and associated for the purpose of publishing their works. The authors who write for the periodicals of these societies are not paid, and the periodicals are not for sale; they are sent gratis to all quarters of the globe, to other societies, cultivating the same branches of learning.

This member of the society may insert in its review a one-page note summarizing his observations; another may publish therein an extensive work, the results of long years of study; while others will confine themselves to consulting the review as a startingpoint for further research. It does not matter: all these authors and readers are associated for the production of works in which all of them take an interest. It is true that a learned society, like the individual author, goes to a printing office where workmen are engaged to do the printing. Nowadays, those who belong to the learned societies despise manual labour; which indeed is carried on under very bad conditions; but a community which would give a generous philosophic and *scientific* education to all its members, would know how to organize manual labour in such a way that it would be the pride of humanity. Its learned societies would become associations of explorers, lovers of science, and workers—all knowing a manual trade and all interested in science.

If, for example, the society is studying geology, all will contribute to the exploration of the earth's strata; each member will take his share in research, and ten thousand observers where we have now only a hundred, will do more in a year than we can do in twenty years. And when their works are to be published, ten thousand men and women, skilled in different trades, will be ready to draw maps, engrave designs, compose, and print the books. With gladness will they give their leisure—in summer to exploration in winter to indoor work And when their works appear, they will find not only a hundred, but ten thousand readers interested in their common work.

This is the direction in which progress is already moving. Even to-day, when England felt the need of a complete dictionary of the English language, the birth of a Littré, who would devote his life to this work, was not waited for. Volunteers were appealed to, and a thousand men offered their services, spontaneously and gratuitously, to ransack the libraries, to take notes, and to accomplish in a few years a work which one man could not complete in his lifetime. In all branches of human intelligence the same spirit is breaking forth, and we should have a very limited knowledge of humanity could we not guess that the future is announcing itself in such tentative co-operation, which is gradually taking the place of individual work. For this dictionary to be a really collective work, it would have required that many volunteer authors, printers and printers' readers should have worked in common; but something in this direction is done already in the Socialist Press, which offers us examples of manual and intellectual work combined.

It happens in our newspapers that a Socialist author composes in lead his own article. True, such attempts are rare, but they indicate in which direction evolution is going. They show the road of liberty. In future, when a man will have something useful to say-a word that goes beyond the thoughts of his century, he will not have to look for an editor who might advance the necessary capital. He will look for collaborators among those who know the printing trade, and who approve the idea of his new work. Literature and journalism will cease to be a means of money-making and living at the cost of others. But is there any one who knows literature and journalism from within, and who does not ardently desire that literature should at last be able to free itself from those who formerly protected it, and who now exploit it, and from the multitude which with rare exceptions pays it in proportion to its mediocrity, or to the ease with which it adapts itself to the bad taste of the greater number? Letters and science will only take their proper place in the work of human development when, freed from all mercenary bondage, they will be exclusively cultivated by those that love them, and for those that love them.

GENERAL PRINCIPLES

Production of genetically pure and otherwise good quality pedigree seed is an exacting task requiring high technical skills and comparatively heavy financial investment.

During seed production strict attention must be given to the maintenance of genetic purity and other qualities of seeds in order to exploit the full dividends sought to be obtained by introduction of new superior crop plant varieties. In other words, seed production must be carried out under standardized and well-organized condition.

GENETIC PRINCIPLE

- *Deterioration of varieties:* Genetic purity (Truness to type) of a variety can deteriorate due to several factor during production cycles. The important factors of apparent and real deterioration of varieties) are as follows:
  - *Developmental variation:* When the seed crops are grown in difficult environment, under different soil and fertility conditions, or different climate conditions, or under different photoperiods, or at different elevation for several consecutive generations, the developmental variation may arise some times as differential growth response. To minimize the opportunity for such shifts to occur in varieties it is advisable to grow them in their areas of adaptation and growing seasons.
  - *Mechanical mixtures*: This is the most important source of variety deterioration during seed production. Mechanical mixtures may often take place at the time of sowing, if more than one variety is sown with same seed drill; through volunteer plants of the same crop in the seed field; or through different varieties grown in adjacent fields. Often the seed produce of all the varieties are kept on same threshing floor, resulting in considerable varietal mixture. To avoid this sort mechanical contamination it would be necessary to rogue the seed fields, and practice the

utmost care during the seed production, harvesting, threshing and further handling

- *Mutations*: This is not a serious factor of varietal deterioration. In the majority of the cases it is difficult to identify or detect minor mutation.
- *Natural crossing*: In sexually propagated crops, natural crossing is another most important source of varietal deterioration due to introgression to genes from unrelated stocks which can only be solved by prevention

Natural Crossing Occurs Due to Following Three Reasons:

- Natural crossing with undesirable types.
- Natural crossing with diseased plants.
- Natural crossing with off- type plants.

Natural crossing occurs due to following factors:

- The breeding system of species
- Isolation systems
- Varietal mass
- Pollinating agent
  - *Minor genetic variations*: Minor genetic variations may exist even in the Varieties appearing phenotypically uniform and homogeneous at the time of their release. During later production cycle some of this variation may be lost because of selective elimination by the environment. To overcome these yields trials are suggested.

    Selective influence of diseases: The selective influence of diseases in varietal deterioration is also of considerable importance. New crop varieties often become susceptible to new races of diseases often caused by obligate parasites and are out of seed programmes. Similarly the vegetatively propagated stocks deteriorate fast if infected by viral, fungal and bacterial diseases. During seed production it is, therefore, very important to produce disease free seeds/stocks.
  - *Techniques of plant breeders*: In certain instances, serious instabilities may occur in varieties due to cytogenetically irregularities not properly assessed in the new varieties prior to their release. Other factors, such as break down in male sterility, certain environmental conditions, and other heritable variations may considerably lower the genetic purity.

Maintenance of Genetic Purity During seed Production

The various steps suggested to maintain varietal purity, are as follows

- Use of approved seed only in seed multiplication.
- Inspection and approval of fields prior to planting.

- Field inspection and approval of growing crops at critical stages for verification of genetic purity, detection of mixtures, weeds, and for freedom from noxious weeds and seed borne diseases etc.
- Sampling and sealing of cleaned lots
- Growing of samples of potentially approved stocks for comparison with authentic stocks.

The various steps suggested for maintaining genetic purity are as follows:

- Providing adequate isolation to prevent contamination by natural crossing or mechanical mixtures
- Rouging of seed fields prior to the stage at which they could contaminate the seed crop.
- Periodic testing of varieties for genetic purity.
- Avoiding genetic shifts by growing crops in areas in their adaptation only.
- Certification of seed crops to maintain genetic purity and quality of seed.
- Adopting the generation system.
- Grow out tests.

AGRONOMIC PRINCIPLES

- *Selection of a Agro-climatic Region*: A crop variety to be grown for seed production in an area must be adapted to the photoperiod and temperature conditions prevailing in that area.
- *Selection of seed plot*: The plot selected for seed crop must be free from volunteer plants, weed plants and have good soil texture and fertility. The soil of the seed plot should be comparatively free from soil borne diseases and insects pests.
- *Isolation of Seed crops*: The seed crop must be isolated from other nearby fields of the same crops and the other contaminating crops as per requirement of the certification standards.
- *Preparation of Land*: Good land preparation helps in improved germination, good stand establishment and destruction of potential weeds. It also aids in water management and good uniform irrigation.
- *Selection of variety*: The variety of seed production must be carefully selected, should possess disease resistance, earliness, grain quality, a higher yielder, and adapted to the agroclimatic conditions of the region.
- *Seed treatment*: Depending upon the requirement the following seed treatment may be given:
  - Chemical seed treatment.
  - Bacterial inoculation for the legumes.
  - Seed treatment for breaking dormancy.

- *Time of planting:* The seed crops should invariably be sown at their normal planting time. Depending upon the incidence of diseases and pests, some adjustments, could be made, if necessary.
- *Seed Rate:* Lower seed rates than usual for raising commercial crop are desirable because they facilitate rouging operations and inspection of seed crops.
- *Method of sowing:* The most efficient and ideal method of sowing is by mechanical drilling.
- *Depth of sowing:* Depth of sowing is extremely important in ensuring good plant stand. Small seeds should usually be planted shallow, but large seeds could be planted a little deeper.
- *Rouging*: Adequate and timely rouging is extremely important in seed production. Rouging in most of the field crops may be done at many of the following stages as per needs of the seed crop.
  a. Vegetative/ preflowering stage
  b. Flowering stage
  c. Maturity stage
  d. Supplementary pollination

Provision of honey bees in hives in close proximity to the seed fields of crops largely cross pollinated by the insects, ensure good seed set thereby greatly increase seed yields.

Weed Control

Good weed control is the basic requirement in producing good quality seed. Weeds may cause contamination of the seed crop, in addition to reduction in yield:

Disease and Insect Control

Successful disease and insect control is another important factor in raising healthy seed crops. Apart from reduction of yield, the quality of seeds from diseased and insect damaged plants is invariably poor.

Nutrition

In the nutrition of seed crops, nitrogen, phosphorus, potassium, and several other elements play an important role for proper development of plants and seed. It is, therefore, advisable to know and identify the nutritional requirements of seed crops and apply adequate fertilizers.

Irrigation

Irrigation can be important at planting for seed crops on dry soils to ensure good uniform germination and adequate crop stands. Excess moisture or prolonged drought adversely affects germination and frequently results in poor crop stands.

Harvesting of Seed Crops

It is of great importance to harvest a seed crop at the time that will allow both the maximum yield and the best quality seed.

Drying of Seeds

In order to preserve seed viability and vigour it is necessary to dry seeds to safe moisture content levels.

Storage of Naw Seeds

The best method of sowing seed for short periods is in sacks or bags in ordinary buildings or godowns.

SUNFLOWER TECHNICALITY

Sunflower scientifically known as *Helianthus Annuus* is an annual plant that belongs to the family of *Asteraceae* and is native to North and South America. Although it is not commonly grown in the Philippines, it can thrive in its soil. The common and recommended variety of sunflower in the Philippines is the hybrid type that offers many uses which will be sourced out from Central Luzon State University. Its seeds can produce high-value oil that can be used in various food and product manufacturing. Its oil is predominantly oleic acid that is widely-used in food industry. The edible oil can lower blood levels of cholesterol, improve lipid profiles and modify harmful inflammation. Most importantly, this is now a good source of biofuel through its seeds.

Sunflower has generated a lot of interest among farmers and therefore its potential as an economic crop needs to be evaluated and determined. As sunflower farming has seen its potential into a mainstream industry, the future of the farmer growers and the market for seeds that can be processed into biofuel including sunflower seeds would be guaranteed since Philippine National Oil Company – Alternative Fuel Corporation (PNOC-AFC) and Minola Oil Company would purchase sunflower seeds in commercial quantity for the production of crude oil. Based on the one cropping cycle operation of a one hectare land, the project can generate total sales amounting to Php 40,000.00.

To grow sunflower well, the area for planting should have good irrigation facilities. A moderate to well-drained soil is the basic soil requirement. Soil used in growing corn, rice, and vegetables is also suitable for sunflower farming. It is important to prepare the land before planting sunflowers. Land preparation includes clearing, plowing and harrowing and sunflower seeds production includes planting, application of fertilizer, hilling-up, weeding and maintenance, irrigation/ watering, spraying, harvesting, threshing, drying, cleaning and bagging. This lasts for three (3) to four (4) months. A hectare of land can be planted with 47,620 planting hills. The recommended system of planting is single row with 75 centimeters space between rows and 25 centimeters between mounds. Seeding rate is 5-6 kilograms per hectare given that there are 1-2 seeds with 3-4 centimeters depth for each mound. One cropping cycle can yield a total of 2,000 kilograms of dried sunflower seeds. Luckily, Region XII has favourable climate and large areas suitable for sunflower plantation especially in provinces of South Cotabato and Sarangani where idle lands are mostly found.

Sunflower is grown at minimal cost. The estimated initial investment cost per hectare amounts to Php 33,657.50 with working capital of Php 25,657.50. The Net Income before tax is calculated to reach Php 14,175.83 per cropping cycle. The Gross Profit Rate and Net Profit Rate are 40 per cent and 35 per cent respectively. This undertaking is expected to have a Return on Investment of 126 per cent. Approximately after nine (9) months and eleven (11) days, the cost of investment will be recovered. Thus, the project is worthwhile to realise.

POLITICS OF SEED

John highlighted how much of the debate surrounding Africa's attempts to launch a new Green Revolution stresses a narrow theory of change which emphasises 'market led technology adoption'. This fails to address the underlying political economy of seed policy processes and the different configurations of actors and interests that affect the way cereal seed systems operate.

In particular, it misses out the crucial politics behind the narratives and framings that drive R&D strategies and investment decisions behind this Green Revolution agenda. Drawing on lessons from country case studies from Kenya, Ghana, Malawi and Zimbabwe, Thompson showed that by addressing the politics behind this agenda – and with it, the interests, values and choices that drive innovation – a credible, politically informed debate can be opened up about future options and alternative pathways.

GREEN REVOLUTION

Elijah noted that Kenya is in many ways seen as the 'poster child' for Africa's new Green Revolution, with numerous public-private partnerships promoting a growing network of 'agro-dealers'. He reported however, that he and his co-researcher Hannigton Odame found that despite claims that the agro-dealer model offers the best approach for delivering new seeds and technologies to farmers, reality on the ground has yet to match expectations. Agro-dealers remain spread unevenly across the country and are inevitably concentrated in the higher potential areas The changing structure of Kenya's seed industry and entry of multinational and philanthropic players (who focus on delivery of hybrid maize and fertilizer) are shifting this dynamic further. This approach is acting to narrow the choice of seeds and crop types to many farmers, leading to technological 'lock in' – particularly with in the maize system – to the exclusion of other viable options.

THE POLITICS OF SEED

Douglas explained that a decade of political and economic turmoil, as well as a period of radical land reform, affected Zimbabwe's seed system, reducing the supply of quality seeds and undermining regulatory control.

The collapse of the seed system was exacerbated by seed relief programmes implemented by government, donors and NGOs alike, which bypassed the normal market chain. In 2010, aid agencies experimented with 'market-friendly' input subsidy programmes to promote uptake of improved seeds and fertilizers, but these also created distortions and became objects of political manipulation and contestation. He argued that improving the effectiveness of the input programmes and strengthen the re-emerging seed system will require addressing overlapping objectives, improving coordination and increasing accountability and trust among key stakeholders.

SEEDS IN AFRICA

McCann's talk drew on research from both Ethiopia and Zimbabwe. He observed that agricultural history and the history of seeds in sub-Saharan Africa are an aggregate effect of individual day-to-day decisions by farmers operating in diverse environments. Farmers' seed selections are based on a complex of both natural conditions – *e.g.*, moisture, pests, soils – and short-term, season by season farm decisions about labour, potential yield vs. risk and market potential. In the immediate future, international donors and African governments are planning that African farmers will receive their seeds from a global political structure that anticipates, perhaps wishfully, economic, ecological and political stability.

McCann noted that those expectations of agricultural scientists, policy makers and multinational seed companies are the same ones that failed Africa's farmers at the end of the twentieth century. Thus, he cautioned that the new push to develop and deliver improved hybrid and GM varieties of seed to farmers in complex, diverse, uncertain environments through systems that assume economic, social and political stability may end up repeating the mistakes of the past.

SEEDS AND SUBSIDIES

Chinsinga was a critical account of the cereal seed systems in Malawi, in both a historical and contemporary context, with particular reference to the three input support programmes implemented since the late 1990s to date. He noted that the centrality of the question of food security in the country's electoral politics in a post-liberalisation context has created a seed industry dominated by multinational seed companies, who offer farmers a very limited range of products, mainly imported hybrid maize which is not always suited to local conditions, while weakening the national breeding programme. The interests of seed companies, donors and government have, for different reasons, coincided to create a seed industry that serves a narrow set of interests. This includes a network of private agro-dealers which have been captured by urban elites who often only open during the subsidy season then close during the rest of the year.

At the close of his talk, Chinsinga noted that while the dominant players in Malawi's 'Green Revolution Alliance' have shaped the major policy directions of agriculture in the country, this has not always benefitted smallholder farmers. Thus, he reasoned, once these subsidy programmes come to an end, there is a very good chance that farmers will revert to local maize varieties and other crops and many of the 'suitcase' agro-dealers will go out of business.

SEED MULTIPLICATION

Governments in the region noted about 30 years or so ago that the disparity between food production and population increase in Africa south of the Sahara was ever widening with a resultant decrease in per capita food consumption. Due to population increase land availability was declining. The best approach to increasing food production was, therefore, to increase crop yields. But farmers were growing mostly their own traditional varieties, which are low yielding. Some high yielding varieties had been released but were not available to farmers.

As a result quasi-government seed companies were established in Malawi (National Seed Company of Malawi, NSCM) and Tanzania (TANSEED). After a few years of operation, these companies abandoned their self-pollinating crop seed enterprises citing poor profitability. As a result, low adoption of recommended technologies for beans still remains a major problem in the region despite the development of new high yielding varieties. Such barriers to seed dissemination are a major constraint to the determination of impact from CRSP programs. The East Africa Bean/Cowpea CRSP Project therefore embarked on development of locations and means to multiply and disseminate basic and foundation seed of improved varieties.

Earlier studies at Bunda indicated that growing beans in Malawi during the hot dry season acts as a filtration process of seedborne diseases except viruses. As a result, the programme embarked on increasing breeders' seed under irrigation on campus. In each of the past five years breeders' seed has been increased to supply basic seed. At SUA basic seed is also produced similarly. During the past five years the Malawi Bunda CRSP multiplied basic seed of Kalima on about 5ha in each year to produce foundation seed. This seed was sold to NGOs, *e.g.*, ActionAid and Self-Help Development International, and to Ministry of Agriculture for demonstrations. In Tanzania the bean seed distribution system is significantly different from that in Malawi. Until recently, bean seed was produced exclusively on government farms with little seed being produced or reaching the farmer. As the shortcomings of the official Ministry of Agriculture multiplication and dissemination stations become apparent and as market liberalization has progressed, more decentralized seed production systems are developing.

Barriers to seed dissemination result because of a number of socio-economic constraints. Poor distribution of inputs and produce in the region results from poor infrastructure that exists. The road/bridge network is so vulnerable that vast areas are inaccessible particularly during the rainy season. Markets are not adequately established. The quasi-government marketing organizations have better networks of markets but are poorly organized. They offer the farmers the lowest prices and generally do not have ready cash to offer the farmer when the crop is ready for sale. Entrepreneurs are coming into the system due to market liberalization. They may have ready cash and transport but lack necessary knowledge to handle seeds differently from other inputs and produce. In many instances farmers are not organized into groups or cooperative societies, which are powerful bodies that would assist them do combined efforts to break through the barriers of the marketing systems. In Malawi the credits system broke down as the country changed from dictatorship to a plural society.

It has been realized that smallscale bean farmers will:

- Buy bean seed and not just rely on their own stocks,
- Afford to buy seed of new varieties, and
- Not function efficiently in varietal diffusion.

Non-governmental organizations are increasingly filling a niche in seed dissemination activities and thereby providing temporary solutions to the barriers expressed previously and to facilitate farmers' capabilities for seed production. Their usual goals are food security and poverty alleviation. Malawi has a greater profusion of NGOs because of its burden of Mozambican refugees in the 80's and 90's. These NGOs have stayed even after most refugees were repatriated. In general, these NGOs operate two types of seed schemes; *viz.*: commercially-oriented schemes where fewer farmers purchase more seed to produce seed for sale and the food security oriented schemes where more farmers are given less seed on credit to produce seed or grain for the market or food. The former crop is certified whereas the latter is hardly inspected but the seed is 'approved'. Approved seed is that seed which is produced either from certified or known stocks of uncertified seed. Seeds of this category do not necessarily carry a government seal but have to pass seed standards for purity and germination. This has facilitated the operations of many NGOs in seed multiplication and dissemination in Malawi. In Malawi two organizations involved in seed distribution are described. The first is an NGO (ActionAid) that has a food security-oriented seed scheme. ActionAid developed a 'Malawi Smallholder Seed Development Project, MSSDP' after a flood disaster in 1992 when 1.1 million households benefited from 3,000 tonnes of relief seed issues.

The objectives of this Project are to:

- Develop low cost mechanisms, which would improve and sustain availability and accessibility of appropriate, improved seed to resource poor farmers in participating communities,
- Establish sustainable self-motivating community based groups, which will manage seed multiplication and distribution in the communities, and
- Train selected extension staff and participating community groups in seed production, seed quality control, on farm seed selection and storage, group dynamics and in community participatory methodologies.

In 1998 a total of 3.6 tonnes of certified bean seed was supplied and made available to 76 groups with a total membership of 1500, who in turn multiplied more than 12 tonnes of first generation approved bean seed. About 75 per cent of this seed were Kalima and Nasaka from CRSP.

An example of the commercially oriented seed scheme is that operated by the Maize Productivity Task Force, formed in Malawi in 1995, with the aim of increasing agricultural productivity. Its Action Group Two, one of the four groups, was set up to increase availability of seeds of self-pollinating and open pollinated crop varieties, which are neglected by the large multilateral seed companies. Funded by European Union, this group embarked on seed production and marketing as a business.

The major activities of this group are:

- Technology promotion through demonstrations, radio programs and field days;
- Training of extension staff and seed growers in seed production techniques;
- Organization of farmers through establishment of farmer associations; and
- Seed production.

In 1998/99 season there were 109 farmers growing seed beans on 28ha producing 24 tonnes of seed. The CRSP project in Tanzania released SUA90 and Rojo. These varieties have been introduced into some selected farming communities in the country. The areas where they are currently grown and marketed include Morogoro and Kilosa districts of Morogoro region, Kongwa district of Dodoma region and some areas of Tanga and Arusha regions. Seed of Rojo was recently produced at SUA, Tanzania and was purposely disseminated through two NGOs, *viz.*: LVIA (Lay Volunteers International Agency) and CCT (Christian Council of Tanzania) and through the CRSP Project. The CRSP Project distributed the seed to Kongwa, Kisanga and

Msolwa villages in addition to farmers in Maharaka and Msongozi, Muhenda and Ulaya-Mbuyuni villages of Morogoro and Kilosa districts. LVIA selected participating farmers based on their willingness to engage in seed production and on financial ability to start seed production. For SUA CRSP gender and education were major criteria. CCT selected progressive and innovative farmers in terms of knowledge on improved agricultural production. Notable amongst the results is the fact that smallholder farmers are willing to purchase foundation seed. However, seed demand and marketing issues are clearly not addressed. Farmers were unable to sell their seeds within their villages and had to sell as cheap grain. Adequate extension is lacking. The majority of farmers are still unaware of the new seed. LVIA provided its own extension staff and farmers under their scheme were more knowledgeable. Seed promotion activities also worked better for farmers under their scheme. A general recommendation was that successful seed producers are those more financially endowed and that these are the farmers to be targeted in future.

SEED SYSTEMS IN DEVELOPING COUNTRIES

The formal seed sector is an official or private control of seed monitored through the entire process of breeding, multiplication, processing and storage, leading to the final product.

The informal seed sector is simply the farmers themselves that provide each other and themselves with seed for sowing. This seed may be cleaned manually, but is otherwise untreated and thus a potential carrier of various diseases.

In industrialized countries, the formal seed sector provides the vast majority of seed to farmers, while both seed systems are present in developing countries. Despite large investments in formal seed systems in developing countries over the past 30 years, the vast majority, about 90–95 per cent, of smallholder farmers' seed demands are still met by informal sources at farm and community level.

Although the informal seed sector provides a dynamic and flexible system of seed supply, usage, handling, trade and exchange, continuous use of untested seed inevitably leads to a degeneration of the seed quality - genetically, physiologically and seed health wise. Farmers depend on their own seed for sowing, not only because of the inadequate access to seeds from the formal seed sector, but also because the formal seed sector more often provides seeds of a limited range of cultivars and varieties, which not always fulfil the needs of the farmers to diversify the production. In addition, the high cost of these seeds quite often make them unattainable for the poor farmers.

On-farm growing and maintenance of locally adapted landraces, cultivars and wild species help the farmer decrease the impact of a series of production constraints like drough, tflooding, heat, cold, pests and diseases. In many developing countries, problems created by seed-borne diseases are ignored and control measures unknown or inadequate. The consequence is often poor seed quality, dissemination and build-up of seed-borne diseases and yields far below the potential.

The quality of the seed must be known before it is sown. A farmer using only healthy seed will be able to increase the yield and quality of his harvest dramatically. However, the health and quality of a seed is not always apparent to the naked eye.Seed supply from both the formal and informal systems suffers from a series of problems due to the lack of economic resources for education, research and quality control.

The Following Problems are Addressed at DSHC:

- Low quality of seeds
- Limited use of clean and healthy seed by farmers and seed producers
- Problems of seed health in crop improvement programmes
- Indiscriminate use of pesticides as seed treatment
- Lack of trained manpower in seed pathology
- Lack of facilities and seed health testing equipment
- Lack of knowledge on the significance of seed-borne infections in the field as compared to other means of transmission *e.g.*, soil and collateral hosts
- Insufficient transfer of technologies suitable for subsistence farmers

Small-scale seed production in certain developing countries deals with the "quality declared seed system of FAO/ SIDA". Standards are established or adjusted according to countries. The quality of seed is to be controlled by the producer. The producer declares the quality of the seed, while the governmental local seed controlling agencies only control a part of the seed production (approx. 10 per cent) with random control like spot-checking and field inspections. It is suggested that private laboratories and seed technologists nominated by government agencies should support the local extension services to guide the farmers' production and still may have the mandate to test breeders' basic and quality declared seed.

MAIZE SEED SYSTEMS

As increasing numbers of genetically modified crops are released into the environment, the likelihood of unintended ecological effects on both agricultural and natural systems increases. These effects become particularly relevant in centers of crop origin and diversity. In Mexico, a country that harbors over 60 per cent of maize's genetic variation, gene flow among landrace and teosinte populations has occurred readily since maize's domestication 9,000 years ago. But unlike domestication genes, which often represent a loss of function that decreases a plants' ability to survive without human intervention, many transgenes (*e.g.*, Cry genes) represent a gain of function that could enhance the survival or even the weediness of wild relatives.

Assessing the potential for the dispersal of transgenes into crop landrace and wild populations is critical. The presence of transgenes in Mexican maize landraces was first reported in 2001 in the state of Oaxaca, but the extent of their dispersal is still in question. A subsequent study reported the presence of transgenes, while a third failed to detect them. Some suggested that transgenes had disappeared, but recent studies have confirmed their presence in Oaxaca and found them in a new area of Mexico.

Inconsistencies across studies might be due to differences in the analytical methods used or to narrow geographic sampling. Most analyses to date have been based on haphazard sampling of fields and seed stocks in a restricted number of localities; results are not representative of a well-defined population. Discrepancies might also be due to the dynamics of seed populations. However, the absence of proper data on seed dynamics and a formal framework to interpret these data has lead to widespread speculation.

In this paper, we analyse the implications of seed dynamics on the dispersal of maize transgenes across Mexico. There have been no commercial releases of genetically modified varieties (GMVs) of maize in Mexico, and there was a moratorium on all open-field plantings after 1998. However, seed of maize GMVs can be purchased in the United States (US), where it is widely planted, and brought into Mexico. US maize grain is another possible source of transgenes, since millions of tons of non-segregated grain have been imported and distributed throughout Mexican rural areas by the public retail network Diconsa. Seed and pollen exchange are both essential for the dispersal and persistence of alleles in cross-pollinated plants, yet there has been scant research on the effect of seed exchange on crop genetics.

Current models of transgene dispersal focus almost exclusively on pollen exchange and the selective advantage of transgenes in wild populations. Although they are well suited to industrialized agriculture, where seed is an input replaced every cropping cycle and seed exchange is absent, these models are not appropriate wherever seed is a capital asset saved across cropping cycles. In most centers of crop diversity, including Mexico, farmers save seed across cycles, forming local seed stocks, and they exchange seed among each other creating informal seed systems. Seed systems consist of an interrelated set of components including breeding, management, replacement and distribution of seed. In addition to seed systems, farmers occasionally use grain purchased as food or feed in lieu of seed. Although there have been recent attempts to model the role of seed movement and anthropogenic factors in the establishment of feral crop populations and volunteers in industrialized agriculture, seed dynamics in centers of crop diversity constitute an entirely different phenomenon. In contrast to pollen, which deposits largely within meters, seed and grain can move thousands of kilometers, and seed replacement can alter local allele frequencies instantly and decisively. Unsurprisingly, some analysts have assumed that maize germplasm introduced into Mexico, including GMVs, can diffuse rapidly across the country through informal seed systems and grain markets.

It is undeniable that genes can linger in or travel across local seed stocks as a result of farmers' decisions, but there are no quantitative analyses of this process. Here, we assess the potential for transgene dispersal via seed based on a model of crop populations and nationwide data on maize seed management. We assess the distribution of transgenes across Mexico, and we test whether this distribution can be explained through different combinations of previously proposed mechanisms.

MATERIALS AND METHODS

Studies of transgene dispersal face several methodological challenges. Although both polymerase chain reaction (PCR) and enzyme-linked immuno sorbent assays (ELISA) offer reasonable accuracy in the detection of transgene frequencies above 0.5 per cent, frequency estimates themselves are still problematic. Quantitative estimates often depend on the screening method used. They also depend critically on the sampling framework. Even when transgene frequencies in a sample can be determined with reasonable accuracy, inferences on their frequency in the field must account for the structure and dynamics of the crop's metapopulation.

Gene frequencies are scale-dependent due to the influence of population structure on gene flow. Spatial structure determines pollen exchange within and among individual plots in a locality during a single cropping cycle, but seed dynamics and management can have an overwhelming influence on the structure of populations across cycles and locations. It is misleading to estimate allele frequencies beyond the plot level without unraveling this complex population structure.

In order to avoid these shortcomings, we focus here on the dynamics of relatively homogeneous populations, *i.e.*, seed lots, and the presence/absence of recombinant proteins within them. A seed lot is defined here as the set of kernels of a specific type (*e.g.*, shape, size or colour) selected by a farmer and sown during a cropping cycle. A transgenic seed lot is defined as one which contains one or more seeds expressing recombinant proteins. Seed lots and groups of seed lots that share some characteristic (*e.g.*, origin) are often subject to distinct rates of replacement and diffusion, which means that they constitute a well-defined seed population that can decrease or increase in numbers within the crop's metapopulation as a function of seed management. Thus, the dispersal of genes within and across crop populations can be fostered or strictly limited by farmers' management practices.

The rate of growth (λ) of a closed seed population depends on the rates at which farmers save seed across cycles (p) and diffuse it (q) among a number (C) of fellow farmers: $\lambda = p + qC$. In general, seed type i will grow as long as $\lambda i>1$. Seed that is not saved must be replaced, so that the rate of seed replacement is equal to $1-p$. In a metapopulation of constant size, a seed type that exhibits higher rates of replacement or lower rates of diffusion than the rest will decrease until it becomes extinct. In 2002, the total maize acreage in Mexico was constant relative to previous years, and the estimated growth rate of the landrace metapopulation was $\lambda = 1.03$. Hence, differences in the rates of replacement and diffusion across maize seed types will indicate their propensity to spread within the metapopulation.

We estimated the frequency of presumed sources of maize transgenes and the rates of seed replacement and diffusion using data from the nationally representative 2002 Mexico Rural Household Survey. This allowed us to analyse the presumed mechanisms of transgene dispersal into landrace populations. Using ELISA, we screened a collection of all maize seed types kept by survey households to determine the presence of transgenes. We tested for activity of two specific recombinant proteins from the most common commercial maize GMVs in the US in 2002: CP4/EPSPS and Cry1Ab/Ac. While PCR is perhaps the most common transgene detection method, ELISA's accuracy in qualitative analysis is comparable.

ELISA has been thoroughly validated for transgene detection in maize it offers clear advantages when screening large samples and is widely used in scientific research. By screening directly for active recombinant proteins, we avoid technical problems associated with establishing the presence of recombinant DNA sequences from leaf tissue. Our frequency estimates might be conservative if transgenes are present but inactive due to silencing and given that we screen for only the most common recombinant proteins.

SEED-LOT SAMPLE AND SURVEY DATA

The survey is representative of the rural population nationwide and in each of the five regions in which INEGI divides the country. It is based on a stratified, three-stage cluster sampling frame designed by INEGI. Within each region, a sample of states, localities and households (*i.e.*, primary, secondary and elementary sample units, respectively) was selected through simple random sampling at every stage. Hence, our household sample consisted of 1765 households in 80 localities across 14 of the country's 31 states. The survey provides detailed information on the activities and assets of the rural population. ENHRUM also gathered data on every maize seed lot (*i.e.*, every distinct seed type) managed by households at the time of the survey, including detailed data for 2002 and retrospective data on seed diffusion for the previous 5 years. Data on 861 maize seed lots from 606 households were used to estimate rates of seed replacement and diffusion.

Since these data are derived from a census of seed lots owned by surveyed households (*i.e.*, there was no sampling of seed lots within households), there are no sample design effects to consider other than those pertaining to the sampling of households themselves. As with most surveys, the precision of variance estimates derived from ENHRUM data is affected by its complex sample design. While clustering increases the variance of estimates, stratification entails a gain in precision of 21 per cent relative to simple random sampling. Although it is possible to correct for design effects on the variance of simple descriptive statistics (*e.g.*, means and aggregates), no correction methods are available for most analytical statistics, including the ones presented in this paper, which assume a simple random sampling of households.

SEED SAMPLE AND MOLECULAR ANALYSIS

Survey households also provided three seed-quality maize ears (mazorcas buenas para semilla) of every type they owned. This entailed selection out of seed stocks (or a harvest pile) according to farmers' criteria, which tends to sort out unintended crosses exposed by xenia when the pollen's genotype has a visible influence on the development of the endosperm. A total of 419 seed lots were collected from 286 households in 49 localities across the 14 states. Seed replaced or discarded by households after the 2002 harvest was surveyed but not collected. Hence, the collection is representative of seed stocks at the beginning of 2003, which allows us to assess transgene dispersal up to the summer/fall 2002 cycle. Despite a larger sampling effort in the northeast and northwest, little maize was collected in those regions because commercial seed, which is common there, is replaced annually. The two regions are treated here as one. Seed is stored at El Colegio de la Frontera Sur (ECOSUR) and identified by blind-code collection numbers.

In order to maximize the number of pollination events sampled, two complete rows were removed from every ear in the ENHRUM collection and sown in bio-controlled greenhouse conditions until the six-leaf stage. Leaf tissue of 20 randomly-chosen individuals per ear was then pooled to integrate a single sample for each seed lot. Our protocol entails a sample size (n) of 60 seeds per lot, allowing detection of transgenic seed frequencies >0.045 (*i.e.*, >4.5 per cent) at P<0.05. This corresponds to GMV seed lots and some advanced-generation seed mixtures resulting from different combinations of crossing; *e.g.*, selfing of GMV×non-GMV hybrid or backcrossing and reciprocals of a GMV×non-GMV cross with a non-GMV. In some cases, <60 seedlings per lot reached the six-leaf stage, reducing our ability to detect transgenes. Commercial DAS-ELISA kits used can detect 1 seed expressing CP4/EPSPS in 1000 and 1 leaf in 100. A test of 1750 seed and leaf samples expressing CP4/EPSPS and 1750 conventional EPSPS samples, performed by the manufacturer, showed no false positives or negatives. We performed duplicate tests for each sample to increase the reliability of results. In order to avoid contamination, tissue samples and controls were processed separately according to the standard protocol. Kits were used on duplicate tests of 327 samples (10,979 individual seedlings) for CP4/EPSPS and 321 samples (10,679 seedlings) for Cry1Ab/Ac.

As a negative control for both assays, we used leaf tissue of glufosinate resistant maize from the biolistic transformation of the CML72×CML216 hybrid introducing the pat gene (encoding phosphinothricin-N- acetyltransferase). CML72 and CML216 are two of CIMMYT's tropical inbred maize lines. Leaves of maize plants expressing Cry1Ab/Ac and CP4/EPSPS were used as positive controls. Optical density (OD) was measured at 650 nm in a spectrophotometer after incubating for 10 minutes. Positive controls showed readings equal to the positive lyophilized protein provided with the kit. Negative controls (CML72, CML216) were consistently non-reactive to CP4/ EPSPS and Cry1Ab/1Ac. Positive threshold values (Th) were defined as OD mean+5 SD of the normalized blank and negative control leaf tissue values, which is a more stringent criterion than the manufacturer's. Thresholds were set to ThCP4 = 0.154 for CP4/EPSPS and ThCry = 0.142 for Cry1Ab/Ac. Only samples with duplicate positive measurements (above the threshold) were considered positive.

Analytical results were used to estimate frequencies of seed lots containing transgenes at the regional and national level but not the frequencies of transgenes within seed lots. A focus on presence/absence of transgenes at the seed lot level is entirely compatible with our interest in long-distance dispersal via seed. We have analysed transgene dispersal at the locality level using a very different methodology and report our findings elsewhere.

RESPONSE TO INFORMAL SEED SECTOR

Farmers grow many varieties of rice. More often than not they use their own seed. Farmers have produced their own seed for generations. Farmers have exchanged seeds with neighbours for generations. This farmer seed system is also called the informal seed sector. The farmer seed system is the system that farmers use to obtain, produce, conserve, improve, and distribute seed. The farmer seed systems are important in both good years and bad, and is also important for maintaining biodiversity. In a farmer seed system, farmers do not usually distinguish between seed and grain. Seed is not grown in separate plots but simply selected from harvested grain. Its quality can vary from very good to very bad. In the very bad, there is diseased seed, mixed seed, and weed seed. Local varieties may only be produced by the farmer seed system and so may continue to be eaten in the village due to the existence of the farmer seed system. In a village, there may be a few farmers who are recognized as good seed growers. Other farmers may receive some of their seed from these farmers. These farmers may also be looked to when a new variety is assessed. In earlier modules, you have learned how farmers can improve the quality of their own seed.

There is the formal seed sector for rice, which includes the government varietal research institutes, government seed production farms, the seed certification agency of the government, committees for varietal release, private companies, and nongovernment organizations with seed production programs and cooperatives registered by the government for seed. In the formal sector, the government develops the rules that regulate the work of each organization. An example of regulations guiding the production and distribution of seed is the Bangladesh Seed Act 1997. The farmer seed system existed way before the formal seed sector.

You will have learned under module 1 that rice is a low value, self-pollinated crop. Rice seed can be produced relatively easily and stored on-farm over a number of years. It is therefore difficult to create a stable demand for external seed supply. Because of this, private companies have not shown a high interest in producing rice seed. Therefore, the formal seed sector has tended to be dominated by the government. The farmer seed system and the formal seed sector exist alongside each other. A new variety may come through the formal seed sector and then continue to be produced each year in the farmer seed system.

RESEARCH

As calls for a 'Uniquely African Green Revolution' gain momentum, the focus on seeds and seed systems is rising up the policy agenda. Much of the debate emphasises the technological or market dimensions, with substantial investments being made in seed improvement and the development of both public and private sector delivery systems. But there is currently much less emphasis on the wider policy dimensions – and particularly the political economy of policymaking in diverse African contexts.

Experience tells us that it is these factors that often make or break even the best designed and most well intentioned intervention. And since investment in seed improvement and supply was last emphasised as a major development priority (in the 1970s and 80s), contexts have changed. The collapse of national public sector breeding systems has been dramatic, and this has only been selectively compensated for by the entry of the private sector. Large multinational seed and agricultural supply companies are increasingly dominating the global scene, and there are many claims made about the promises of new technologies (notably transgenics) transforming the seed sector through a technological revolution. While informal breeding and seed supply systems continue to exist, and indeed have been extensively supported through NGO and other projects, they are often under pressure, as drought, corruption and conflict take their toll and economic transformation and livelihood change continues apace.

The focus on *cereal* seed systems allows the research to concentrate on a similar set of crops across the four study countries with a key influence on food security at household and national levels.

Given the political reverberations of the 'food crisis' of 2007-08, this allows for a timely analysis of the implications of the policy processes shaping the breeding, production, marketing and distribution of cereal seeds. Whether grown for local subsistence or traded commercially, the significance of cereal crops to national politics (and so arguments about food security and sovereignty), commercial interests and local livelihoods – is likely to be profound.

HYPOTHESIS

The project will test the hypothesis that contrasting politics and different configurations of interests will make a difference to the way cereal seed systems operate and how a 'new green revolution' push in envisaged and ultimately plays out. The underlying implication is that politics matter and engagement with policy processes is important – defining and then deliberating among different framings and interests – *i.e.*, beyond the technical/ market fix.

A FOCUS ON SEED POLICY PROCESSES

- How do seed policies get created, by whom?
- How do ideas about what makes a 'good' seed policy evolve and change?
- How are boundaries drawn around seed problems and policy storylines elaborated?
- Whose voices and views are taken into account in the seed policy process? And what/who is excluded?
- What spaces exist for new ideas, actors, networks? How can these be opened up?

METHODOLOGY

Overall, an historical approach will be necessary to trace changes in the way policies have been framed, looking at the shifts in narratives about what the problem is and what should be done about it over time. Changes in the configuration of actors, their networks and associated interests will also help illuminate how contemporary policies have emerged. A basic mapping of the current situation will take place, involving interviews with key players (from government policymakers to public/private, national/ international researchers to commercial sector seed suppliers and traders to farmers in different parts of the country and with different resource endowments).

This approach will allow the research to elaborate (in largely qualitative terms), first, the set of 'narratives' (stories about the problems and the appropriate solutions) being deployed by different people. Second, the way such actors interact and relate will be mapped, highlighting key gaps and connections. Third, the interests of different groupings will be analysed, looking at the competing power relations involved, and asking who wins, and who loses in policy formulation and its implementation. Finally, areas of contention and debate will be identified for each country setting, highlighting areas for institutional and policy development (for example, around issues of regulation, certification, priority setting and so on).

9

Laboratory Methods of Seed

DRY WEIGHT

The most generally accepted measure of maturity is the time when the seed has reached its maximum dry weight, a point called physiological maturity. This means that nutrients are no longer flowing into the seed from the mother tree. Maximum freshweight does not indicate physiological maturity because the maturing seed begins losing water while nutrients are still being accumulated and biochemical processes are continuing. Recurrent dry-weight determination of a series of seed samples can be made and the results extrapolated to the remaining crop, but this method is slow and therefore seldom used.

CHEMICAL ANALYSIS

Biochemical changes take place as seeds mature but relatively little is known for most species. Chemical indices of seed maturity have been determined in a few species *e.g.* content of crude fat and protein-nitrogen, which increase five and four times respectively from immaturity to physiological maturity, are the best chemical indices for Fraxinus pennsylvanica.

But they have no advantages over an examination of the embryo and the colour change of the fruit, and the extra trouble to perfom the analyses does not appear to be justified. Rediske found that Pseudotsuga seeds were physiologically mature when the content of reducing sugars dropped to 14 mg/g.

X-RAY RADIOGRAPHY

The examination of the development of the embryo and endosperm of sample seeds by means of X-ray radiographs is a quick and relatively straightforward method of assessing seed maturity, provided that suitable facilities and skilled technical staff are available. The technique has been used successfully for Tectona and a number of other tropical species, as well as for temperate species such as Pinus strobus. It has the disadvantage of requiring relatively expensive equipment and relies heavily on the judgment of the seed analyst for reliable results.

MOISTURE CONTENT OF FRUITS

Water loss of maturing cones and fruits occurs in many species and is closely related to the maturity of the seed. Seeds of Picea glauca are considered ripe when moisture content falls below 48 per cent, of Larix decidua at 25–30 per cent and of Pinus sylvestris when it falls to 43–45 per cent (fresh weight basis). However, determination of moisture content by drying in an oven suffers the same disadvantage of slowness as determination of dry weight.

FIELD METHODS

SPECIFIC GRAVITY OF FRUITS

As moisture content of fruits and cones decreases with maturation, so does specific gravity or density, the ratio of unit weight to unit volume, decrease. Unlike moisture content, it is not too difficult to determine approximate specific gravity in the field by flotation in liquids of known specific gravity. Specific gravity indices of maturity have been established for cones of a number of coniferous species, and the cone to be tested is placed in a liquid in which it will float if mature and sink if immature. Various mixtures of kerosene (SG = 0.80), light SAE 20 motor oil (SG = 0.88) and linseed oil (SG = 0.93) have been used to prepare flotation liquids having a designated specific gravity. Tests must be made immediately after cones are picked from a tree. Specific gravity indices have proved reliable for some temperate conifers *e.g.* an S.G. of 0.74 for Picea glauca, but not for several southern hardwoods in the USA.

EXAMINATION OF SEED CONTENTS

Examination of seed contents exposed by cutting open fruits or cones lengthwise can be a reliable and simple method of assessing seed ripeness, provided the operator is experienced. Most embryos and endosperm pass through an immature "milk" stage, followed by a "dough" stage when the tissue becomes more firm. Mature seeds have a firm white endosperm (where present) and a fully developed firm embryo.

COLOUR OF FRUITS OR CONES

Colour changes in fruit or cone provide a simple and, in some species, reliable criterion for judging seed maturity, but the operator must be experienced in the characteristics of the species concerned. In common with the specific gravity method, it involves no destruction of the seeds in the sample examined. Colour changes are usually from the green of the immature fruit or cone to various shades of yellow, brown or grey, and this may be accompanied by hardening of cone scales or of the pericarp of dehiscent or woody fruits. Since the seed normally matures before the fruit, it is advisable in some species to time collection at an earlier rather than a later stage of the colour change. Colour change was found to be the most reliable indicator of maturity for general practice in several southern hardwoods in the USA. It has also given good results in a number of temperate conifers. In Malaysia Tamari found that the best results were obtained by timing collection of Dipterocarp fruits when the wings turned brown but before the fruit itself changed colour.

In Thailand cone colour is used as a guide to optimum time of collection of pines, but differs according to species. In Pinus kesiya collection starts when cones have hardened and the colour is changing from green to brown in proportions of 50: 50. In Pinus merkusii optimum time of collection is reached when the majority of cones are brownish and some have started to open (Granhof 1975). Trials with the Zambales provenance of P. merkusii have shown not only that extraction is a much more lengthy and expensive operation with green than with brown cones but also that the seed extracted has lower germination rate. Experience with P. caribaea in Honduras is similar.

Abscission and shedding of fruits is usually a sign of fruit maturity and it might be assumed that it also indicates a high content of sound, mature seeds. This is not always the case. The first seeds or fruits which fall naturally are often of poor quality, in which case it is advisable to reject them and to postpone collection until the peak and the latter half of the season. In Thailand fruits of Tectona grandis start to be shed in March but observations have shown that the most viable fruits are the last to be shed, so collection starting only in April is recommended. The first fruits of Dipterocarp species that fall upon ripening are usually defective and collecting should be delayed until the greater portion of the fruit has fallen.

COLLECTION OF IMMATURE SEEDS

It is general practice to collect seeds when they are mature, because they have a higher germinative energy and a greater longevity in storage than immature seeds. An alternative method is to collect fruits prior to ripening and to store them in relatively cool, well-ventilated conditions, which permit after ripening of the seeds within the fruit. It has shown promise on a research scale in a number of species. There are several reasons for the interest in developing techniques for artificial ripening. They are:

TO EXTEND THE COLLECTION SEASON

The short period between seed maturity and dispersal may place an excessive demand on the availability of seasonal labour and in some areas unfavourable weather conditions in the collection period may aggravate this situation. Lengthening the period available for collection permits better organization of the collections and allows skilled personnel to pick more of the crop. It can be particularly valuable in research, where a large number of seed lots have to be collected from widely scattered localities. Griffin used the technique in his study of Douglas fir provenances.

TO AVOID DAMAGE TO THE SEED CROP BY INSECTS AND OTHER PESTS

Insects, birds, rodents and other pests frequently damage or destroy seeds and fruits when they have reached maturity. Early collection may be one method of avoiding these losses. Damage and deterioration of seeds is usually worse during the period while they lie on the forest floor than during any other stage in their history; any reduction in this period will improve their subsequent viability and longevity in storage.

TO SALVAGE IMMATURE SEED COLLECTED INADVERTENTLY

Untrained collectors of seeds often begin picking fruits and cones too early in the year before they are fully mature. Artificial ripening provides a method for handling this material.

The development of techniques for after-ripening of immature seeds will require more research, before they can be applied to a wide range of species. However, where a problem of rapid dispersal or of seed pests exists and provided the earliest time for safe collection of immature fruits can be established, such techniques can be very beneficial.

DETERMINING WHICH TREES TO COLLECT FROM

If it can be assumed that the seed collector has received clear directions from the seed user as to which species and provenances and, in some cases, from which stands he is to collect, it is still his responsibility to select the individual trees for collection. Criteria will vary considerably according to whether the collections are large-scale for afforestation projects or small-scale for research purposes.

Identification of species presents no problems in monospecific plantations, but is essential and may be difficult in mixed natural forest, especially where very similar species of the same genus occur mixed together as happens with pines in Mexico and Central America, eucalypts in Australia and dipterocarps in S.E. Asia. Unless identification is certain, it is often advisable to collect herbarium specimens as well as seed.

LARGE-SCALE COLLECTIONS

In large-scale collections emphasis is on collecting as much seed as possible, as quickly and cheaply as possible, rather than on very careful selection of the parent trees. It is, nevertheless, essential to avoid collecting seeds from very poor phenotypes or seeds which prove to be empty or non-viable.

Useful guidelines are listed by Stein et al. (1974), and form the basis of the following:

- Collect seed only from healthy vigorous trees of reasonably good form that are making average or better growth.
- Where possible, collect from mature or nearly mature trees. Overmature trees should be avoided, since seeds from them may be of low viability.
- Avoid isolated trees of naturally cross-pollinating species, since these are likely to be self-pollinated. Seeds are likely to be few, of low viability, and any seedlings produced are frequently weak or malformed.
- Avoid collecting in stands containing numerous poorly formed, excessively limby, off-colour, abnormal or diseased trees.

Often it will be necessary to compromise between seed production and phenotypic appearence. No seed should be collected from excessively coarse-branched, vigorous "wolf" trees, even though they often bear a large crop, while trees of exceptionally good form sometimes bear so little seed that they do not warrant the trouble of collection.

The bulk of seed will come from trees which are "average or better than average" in both form and seed production. Although few studies have been made of the reproductive biology of tropical trees, the occurrence of some species at a very low stocking (less than one every km^2) suggests that these must be naturally self-pollinated. Seed collection from such trees is free from disadvantages associated with collections from isolated trees of natural cross-pollinators.

SMALL-SCALE RESEARCH COLLECTIONS

In small-scale collections for research, selection of trees will depend on the precise objective of the planned research. Provenance research is now receiving a good deal of attention in many countries.

Advice of IUFRO on provenance seed collections includes the following recommendations on collections from individual trees (FAO 1969):

- Collect from not worse than dominant and co-dominant trees of average quality, within "normal" rather than "plus" stands. Collections from superior phenotypes, if made, should be kept separately.
- Collect from a minimum of 10 trees, preferably from 25 to 50 in the stand. If the stand is very variable, increase the number of trees. Record the number of trees and the approximate percentage which they form of the stand.
- Seed trees to be at least seed fall distance apart from each other. A distance of 100 m has been adopted for Pseudotsuga. This is to reduce the risk of collecting from half-sib parents. In Australia a

minimum distance of twice tree height is used as a practical rule of thumb (Boland *et al.* 1980).
- Individual seed trees to be marked.
- Collect equal numbers of cones, fruits or seeds per tree.
- In normal first stage provenance collections, seed from individual trees may be mixed together. If special studies on individual genotypes are to be done, seed from each tree should be kept separate.

SINGLE TREE COLLECTIONS

Foresters are interested in the variation within populations and provenance as well as the variation between them. In the case of exotics, one approach for the introducing countries is to study between provenance differences under local conditions first, and to investigate variation between individuals in the best provenance, by means of progeny trials, only at a later stage after the provenances locally best adapted have been identified. If progeny trials are the object of seed collection, it is essential to keep seed from individual trees separate at all stages of collection, transport, processing, nursery and field planting.

The preservation of the identity of individual trees through the collection and extraction phase often requires considerably more effort than bulking the collections.

If the effort is made, then certain advantages accrue, as listed by Turnbull (1975b):

- It permits biosystematic study of genetic variation both within and between populations. McElwee (1969) states that seed collections mustbe kept separate by individual trees throughout the test from collection to outplanting, as provenance tests are weakened by combining seed within a stand, making it impossible to distinguish between seed source and individual variation. However, in large provenance tests with many individual trees, it may be beyond the resources of the investigator to retain the identity of the parents.
- It is possible to manipulate to equalize the amount of viable seed from each tree in the provenance mix if the seed must be bulked prior to sowing.
- It is not always possible to detect trees bearing hybrid seeds in the field. This is particularly the case with eucalypts. If seedlots are kept separate then, following the raising of small samples from each tree, any showing evidence of hybridization can be eliminated before the main trial is established.

SINGLE CLONE COLLECTIONS

In seed collection in clonal seed orchards, the unit of identity to be kept separate is often the clone rather than the individual ramet. In Zimbabwe the practice of keeping individual clonal seedlots separate has been followed for many years and it is considered that it more than justifies the additional costs and efforts involved over bulked orchard collections.

The advantages can be summarized as follows:

a. Maintaining individual clonal identities at all stages from collections through to storage enables one to act with the minimum of delay on most information as it becomes available. The more obvious areas where action can be taken quickly are when rogueing becomes necessary or when susceptibility to pests or diseases becomes apparent, and the undesirable seedlots can be isolated or discarded.
b. Seedlots can be made up and supplied to suit specific sites, making use of the most recent information on genotype/environment interaction from progeny test analyses.
c. In conjunction with (b) above, and using the test results of each clone, composite seedlots can be made up in such a way as to give equal clonal repre-sentation in the final planting stock, avoiding the dominating effects that some clones frequently have in a bulked seedlot.
d. In general, it increases the options available to the user.

COLLECTIONS FOR CONSERVATION

Collections are also made for the purpose of attempting to conserve the gene pool ex situ either as seed in long-term storage or in planted conservation stands. Because exact knowledge of gene frequencies in the indigenous populations is largely lacking, collecting for gene conservation must be based mainly on commonsense.

Similar methods as for provenance collections are likely to be appropriate, with the exceptions that:

- A somewhat larger number of trees per genepool should be sampled. Estimates are in the range or 50 to 100.
- The sample should be a strictly random one and include poorer than average as well as better than average trees, in order to capture as much as possible of the total genetic variation. The only restriction on this principle is the impossibility of sampling trees which are bearing no seed.
- An additional measure to ensure the fullest possible genetic diversity within the seed collected is to collect in a better than average seed year. The better the seed year, the better the representation of male parents contributing pollen as well as of female parents supplying seed.
- The quantity of seed to be collected from each provenance will usually be larger, because the recommended area for a conservation stand, 10 ha, is much greater than the total area of one provenance in a single provenance trial.

ASSEMBLING RESOURCES

One part of planning is the timely assembly of clear information on the nature and magnitude of the seed collection tasks—number of species and provenances, seed quantities, location of stands, best dates to collect etc., as described above. The other part is to select and assemble the resources needed to do the job. Details of the various resources which may be useful are discussed in subsequent chapters. During the planning stage, the leader of collecting operations needs to check the preparations for field work under the following headings.

ORGANIZATION OF COLLECTING TEAMS

Known or estimated output of collecting teams needs to be related to the quantity of seed, number of stands and length of season, in order to determine the required number and size of teams. For example provenance collections of Pinus kesiya totalling 3000 kg of cones from 16 stands can be completed in 30 days by one team in Thailand. If the full season is estimated at 45 days and there is a demand for a total at 9000 kg of cones from the same 16 stands, two teams will be needed to do the job. If planning can be done sufficiently in advance, it may give an opportunity to train additional climbers if they are needed. It is desirable to have at least one tree-climber on the permanent staff, who can be responsible for looking after climbing equipment and for training new temporary climbers. In the field the climbers should be organized in small teams with a foreman in charge of each. In Honduras a team of 6 pairs (with one climber and one ground assistant or anchorman in each pair) has been found to be a suitable size.

ORGANIZATION OF TRANSPORT

Collecting teams need to cut to a minimum the time spent in moving between one site and the next. Transport must be available where and when it is needed. If necessary, extra vehicles may be temporarily hired. In roadless country, advance arrangements may be needed to employ extra unskilled workers to assist in carrying equipment, tents, etc.

ORGANIZATION OF EQUIPMENT

Choice of equipment will vary greatly according to local conditions. The steeper and less accessible the terrain, the simpler and lighter should be the equipment. Whereas highly mechanised equipment such as tree shakers or mechanised platforms may be appropriate in large seed orchards on flat land, only light portable equipment is practical where natural stands are 4 – 6 hours' walk from the nearest road. Apart from collecting tools, safety clothing, first aid equipment and plenty of bags and sacks should be provided.

ORGANIZATION OF RECORDS

Meticulous recording and labelling is essential to good collecting.

Appropriate labels and forms need to be designed well in advance and printed in adequate numbers..

ORGANIZATION OF PERMITS

These are not normally required for forest services collecting in government forest reserves, but may be needed when collecting on private land, in National Parks and special reserves, or in another country.

Even if formal authority is not needed, it is often advisable to inform local communities of proposed operations in advance.

ORGANIZATION OF SEED EXTRACTION

Arrangements for rapid movement of fruits from collection site to extractory may be needed, involving the advance organization of transport. The seed extractory staff must be advised when to expect the fruits. If some prelimi-nary sun-drying of fruits in the forest is planned, polythene sheeting or tarpaulins will be needed.

DETERMINING THE YEAR FOR SEED COLLECTION

EFFECT OF PERIODICITY

Seed-bearing of many forest trees is rather irregular from year to year. One year with a heavy crop (a "seed year" or "mast year") may be followed by one or several years with a poor seed crop or none at all (Morandini 1962). This habit of periodicity in seeding is an important factor to consider when planning seed collecting operations. The collection of seeds in a good year confers a number of advantages. There can be a high intensity of selection of seed bearers, the cost of collection is lower, due to the concentration of the crop, and the seeds will usually be of higher germinative capacity and will retain their viability longer than those collected in a poor seed year.

Damage from insects affects a smaller proportion of seeds in a good than in a bad seed year. A heavy seed crop usually reflects a previous heavy production of pollen, to which all or most trees in the stand have contributed. Collection in a good seed year therefore conserves a higher proportion of the genetic diversity among male parents than collection in a bad year which follows pollination from only a small number of trees.

Periodicity is well documented for many temperate conifers. For example on average Pinus sylvestris bears an abundant crop every 2 – 3 years and Pseudotsuga menziesii every 4 – 6 years in the UK. Since the period between the good crops is not regular, a general rule of collecting three years' sowing requirements whenever a species bears a heavy cone crop is there recommended by Seal *et al*. Periodicity in tropical species is less well documented. Irregular mast years in Triplochiton have a great influence on the regeneration, or lack of regeneration, of that species, although periodicity in pests and diseases may play as big a part in seed production as periodicity in flowering (Jones 1975). Poor seed years have been recorded in Pinus caribaea and P. oocarpa, and P. merkusii. In other species periodicity is not marked. Tectona grandis has generally good flowering each year, although exceptionally good seed years are observed in some localities on a three or four year cycle. Gmelina arborea starts to seed early, from age 3 in the Philippines to age 7 in Nigeria and usually produces regular abundant crops but poor seed years have been recorded, at least in some provenances.

Pinus kesiya bears abundant crops every year within its indigenous range and as an exotic if planted in the appropriate climate. Cassia siamea,Acacia mearnsii, Cupressus lusitanica and ornamentals such as Delonix regia and Jacaranda mimosaefolia are other species which can be expected to flower and fruit profusely each year. Periodicity may vary considerably between species within the same genus. Among eucalypts,E. grandis, E. saligna and E. camaldulensis usually bear heavy seed crops every two to three years, while E. gomphocephala and E. maculataonly seed heavily at longer intervals. Dipterocarps in Malaysia have heavy seed years at unpredictable intervals of one to six years. Periodicity and flowering patterns of eucalypts can change when they are grown as exotics. Eucalyptus maculata and E. citriodora bear much larger crops more regularly when grown in plantations. Even in good years, flowering may vary substantially from one locality to another. Sometimes individual trees of a stand are on different cycles, some flowering abundantly one year, others in the next (Krugman *et al.* 1974).

COUNTING THE FRUIT CROP

For those species which are known to exhibit periodicity in flowering and fruiting, it is highly desirable to visit the stands to be collected well in advance of the fruiting season, in order to assess in which of them the next seed crop promises to be heavy enough to justify the cost of collection. Too little is known about which external factors have a decisive affect on flowering to allow prediction of future seed crops on the basis of climate. Estimation of the crop can best be done by counting flowers or young fruits on a sample of the trees in the collection stands. Assessment of the abundance of flowering can give a preliminary estimate of the potential seed crop, but may be misleading if there are subsequent severe losses *e.g.* from insects, wind or poor pollination. In Eucalyptus regnans trapping studies indicated that only about 15 per cent of flower buds and 30 per cent of flowers can be expected to develop into mature fruits.

In species such as the pines, which take two years from pollination to ripen their fruits, a count of one year old cones can give a useful indication of the next year's crop (Stein *et al.* 1974), which can be confirmed by further inspection a month or two before collection is due to begin. Recurrent personal inspection of the future crop is the ideal and presents no problem if collection is carried out in seed orchards or easily accessible plantations or natural forest. It may be difficult or impossible for collecting teams working in inaccessible areas or for international expeditions working in several countries. In such cases the team leader may have to rely on reports from an experienced correspondent or on estimates made during the previous year's collections. If reliable local information is not available, a special reconnaissance may be justified in advance of the expensive collecting expedition.

Where the main seed collecting areas are located in less accessible places, it is useful to maintain permanent phenological plots in easily accessible locations within the same forest type, to act as biological indicators. Such plots should be monitored regularly so as to obtain a record of the timing and intensity of flowering and fruiting. The performance of the plots will indicate to collectors the best times to go into the less accessible areas of the district, to check for flowering. The boundaries of a phenological district that can be usefully served by a phenological plot must be defined by experience. In Malaysia, a sample of 86 Dipterocarp trees in an artificially established arboretum at the Forest Research Institute, Kepong, is monitored monthly or fortnightly and the percentage of trees flowering in a month or year is used as an index of Dipterocarp flowering for that month or year. This index gives a fairly reliable indication of Dipterocarp phenology within the State of Selangor (about 8000 sq. km) in which Kepong is located.

Since the phenological plot in Kepong is within 10 minutes' walking distance from the researchers' offices and laboratories, a lot of time and money is saved which would otherwise have to be spent on transportation and organization of field trips. Where it is necessary to count the fruits or cones, binoculars or telescopes are an essential aid. They should be of high optical quality. In binoculars a wide field of view should be combined with only moderate magnification; a minimum of 50 mm aperture and magnification x7 or x8 are suitable. The normal method is to make counts on a representative sample of seed trees dispersed throughout the whole seed source. It is necessary to take the sample from within the stand, because the perimeter trees always fruit more heavily than the trees within the seed source. Counting may be done from the ground or by climbing neighbouring trees. Fruits are counted from one side of the crown only and the number counted is converted to an estimate of the total crop for that sample tree by a correction factor which varies according to the species and the abundance of the crop.

In Czechoslovakia (counting from climbed trees) a factor 1.6 is used for Abies alba, which is constant irrespective of the size of crop, because the clustering of the cones near the top of the tree makes counting in this species reliable. For Picea abies the factor varies according to the average number of cones counted per tree; for 1 – 40 cones it is 1.4, for 41 – 70 cones it is 1.8 and for more than 70 cones it is 2.5. In the UK (counting from the ground) a factor of 4 is used for Pinus, Larix and Pseudotsuga, while for species with very numerous small cones the practice is to scan say one tenth of the crown from one side and use a factor of 20. The number of sample trees used for cone counting varies according to the size of the stand. In the UK five trees are sampled in small seed sources less than 0.5 ha and the number increases progressively to 20 trees for seed sources of more than 4 ha (Seal *et al.* 1965). Czechoslovakia uses a series of plots, in each of which about 5 dominant trees are climbed to scan a further 10 – 15 neighbouring trees. In Tasmania extensive eucalypt seed sources are sampled at the rate of 1 tree per ha.

FRUIT CROP RATING METHODS

The results of cone or fruit counting are applied to the stand and expressed as a numerical value on a scale running from total crop failure to exceptional seed years. With experience it may be possible to define the criteria for an economically collectable crop in quantitative terms *e.g.* for Pinus sylvestris in the UK a minimum of 25 seed trees per ha each bearing at least 300 – 400 cones is specified. For conifers in Arizona and New Mexico Schubert and Pitcher defined "Few" cones per tree as 1 – 20, "Many" as 21 – 160 and "Loaded" as over 160 cones per tree. Quantitative classifications of this type will clearly vary to a great extent according to species, provenance and site conditions. More often use is made of qualitative scoring assessments which rely on the experience of the assessor. In Washington and Oregon five ratings are used, as shown below:

RATING EXPLANATION

		Other than True Firs
5	Heavy	- Good Crop of Cones on All Exposed Crowns of Most Trees.
4	Medium	- Good to Medium Crop on ¾ of Exposed Crown on Most Trees.
3	Light	- Good to Fair Crop on ½ of Exposed Crown of ½ of Trees.
2	Very Light	- Some Cones on Some Trees.
1	Failure	- No Cones to a Few Scattered on a Few Trees.
True Firs:		**(Upper $^1/_3$ of Crown)**
5	Heavy	- Good Crop of Cones on Most Upper Branches of Most Trees.
4	Medium	- Good to Medium Crop on Most Trees.
3	Light	- Few Cones on Many Trees.
2	Very Light	- Few Cones on Scattered Trees.
1	Failure	- Nonexistent

- A rating of 4 or 5 is good prospect for all pickers.
- A rating of 3 has possibilities for more experienced pickers.
- A rating of 1 or 2 is poor prospect for all pickers.

Each year the State Forest Service publishes estimated average cone crop ratings by species and geographic area, for the benefit of individual pickers.

They are based on surveys of a number of different stands and the average for each area is expressed to one place of decimals. For example in 1972, a poor year, the best rating was 2.5 for Tsuga heterophylla in the Western Cascade of Oregon, while 1.0 or total failure was recorded for some species in more than one area. In Tanzania a four class scale is used and seed crop estimates are made twice, one at flowering and the other about a month before seed collection.

The classes are:

- *No Seed Crop*: Trees without Flowers and Fruits.
- *Weak Seed Crop*: Flowering and Medium Size Seed Crop on Free Growing Trees and Trees on Free Borders of Stands.
- Medium Size Seed Crop. Flowering and Very Good Crop on Free Growing Trees and on Free Borders of Stands, Trees within the Stands Bearing Crop at the Top of Crowns.
- Very Good Seed Crop. Flowering and Very Good Crop on Most Trees.

In Sweden forecasts of the cone and seed germinability of Pinus sylvestris and Picea abies have been made annually for about 80 years. Estimates are given separately for the different combinations of latitude (in 1° steps) and altitude (in 100 m steps).

ESTIMATING FULL SEED CONTENT BY CUTTING TEST

The methods described above provide an estimate of the cone or fruit crop. It is necessary to relate this to seed production by examining the contents of a sample of the fruit crop. Fruits may develop normally to maturity whether one or one hundred of the contained ovules have been successfully fertilized and undergone normal development; in parthenocarpous species fruits can mature without containing any sound seeds at all. The number of fruits is therefore not always a good guide to the number of seeds.

The method generally recommended is to cut cones or fruits lengthwise and count the number of seeds which can be seen on one cut surface. Special cone-cutting knives have been designed for this purpose. One or two cone samples from each of 20 to 100 trees in an area are suggested for southern pines in the USA, 5 or 10 cones from each of 10 trees in the UK. Only normal seeds should be counted. Underdeveloped seeds which often occur at the top and base of cones are not included. The number of full seeds counted which denote a good crop varies according to species, for example 6 or more indicates a good crop in Pseudotsuga, 14 or more a good crop in Picea sitchensis. The relationship of the total number of full seeds per cone to number of full seeds exposed on a cut surface is known for some species, for example a factor of x4 or x5 is appropriate for Pseudotsuga in the western USA. The average number of seeds per fruit for many tropical species is not known and needs to be established under local conditions; they vary from one seed per fruit in *e.g.* most Dipterocarps to several hundred per fruit in Anthocephalus. With multiseeded fruits, it is likely that the number of seeds developed will vary in accordance with climate, soil fertility and the age of the parent trees. The first crops borne by young trees nearly always contain fewer sound seeds per fruit than those borne by the same trees when fully mature. Examination of a sample of seeds in the fruit serves the additional purposes of indicating the state of development or maturity of the seeds and the incidence of damage by pests or diseases. A final decision on whether the seed crop is sufficiently heavy to justify collection in a particular year should be dependent on both the rating of the fruit or cone crop and the results of the cutting test for determination of full seed content..

DETERMINING THE BEST DATES FOR COLLECTION

Some species in the tropics carry some ripe seed at all times of the year. Even in these there is often a period of maximum seed production, when collection will be cheapest and seed quality highest. In other species, and especially in the temperate zone with its marked distinction between summer and winter, ripe seed is borne for a limited period, often during the autumn. For many species there is good information on average dates of the seeding season, but these averages may not be sufficiently accurate for planning collection in a particular year. The period between seed maturation and seed dispersal is often short, whereas the effects of climate in a given year may displace the dates of seeding by several weeks from the average. In the temperate zone an early spring and dry summer can cause very early seed ripening, while strong, dry winds cause rapid dispersal of the ripe seeds. Cool, wet weather, on the other hand may delay ripening and dispersal by weeks or months. In the dry tropics there are similar annual variations in the dates of the onset of the dry season and of the rains. It is therefore necessary in each year to check the correct timing of collection by examination of the crop itself.

The reconnaissance of the size of the seed crop, made 1 – 2 months before seed collection will also give some indications as to how the seeds are maturing. Conclusions from it should be twofold *e.g.*: -"Stands A, B and C: Very light crops, not worth collecting this year. Stands X, Y, Z: Good crops, seeds probably mature in 4 weeks' time". A final check on the ripeness of seed must, however, be done at the time of collection.

The humid tropics present special problems because seasonality effects are usually subtle or absent and the period of maximum seed production is uncertain. After detection of flowering in a stand from which it is desired to collect seeds, it is important to carry out periodic reconnaissance to check on the progress of fruit maturation. An efficient schedule of reconnaissance requires prior knowledge of the length of time between anthesis (flower opening/pollination) and fruit maturity. Among Malaysian trees, the period from anthesis to fruit maturity varies from 3 weeks for Pterocymbium javanicum to 11 months for Diospyros maingayi (Ng and Loh 1974). In the exotic Brazil nut, Bertholletia excelsa, the period is 15–16 months. In Malaysia it is recommended that if the maturation period is X weeks, the development of the crop should be checked at 1/2 X and 3/4 X weeks after flowering. A fixed and arbitrary schedule of, say, once a month, will result in the collector being too late for a fast-maturing fruit like Pterocymbium javanicum and too wasteful of effort in the case of a slow-maturing fruit like the Brazil nut.

Apart from the exceptional case of deliberate collection of immature seeds (discussed below), seed collectors need to be able to time collection for the period when seeds (but not necessarily fruits) are fully ripe but before they are dispersed by fruit dehiscence or fruit consumption by animals. To achieve this aim collectors must be able to distinguish ripe from unripe seeds.

Several different methods have been used for the recognition of seed maturity. None of them work perfectly on all species and a good deal of experience or research is needed to determine the best method, or combination of methods, for a hitherto unfamiliar species. They may be divided into those which are of direct application in the field and those which need laboratory equipment. The latter may be valuable in providing a check on the field methods, but are unlikely to be of practical use to the collector unless the collection site happens to be close to the laboratory, which may be the case for some seed orchards.

10

Seed in Flowering Plant

DICOTYLEDON

Dicotyledons or "dicots" is a name for a group of flowering plants whose seed typically contains two embryonic leaves or cotyledons. There are some 199,350 species within this group Flowering plants that are not dicotyledons are monocotyledons, typically having one embryonic leaf. It is now accepted, thanks to the Angiosperm Phylogeny Group, that monocotyledons evolved from within the dicotyledons, and as such the latter form a paraphyletic group. This means that dicotyledons no longer are regarded as a "good" group, and the names "dicotyledons" and "dicots" are no longer to be used, at least in a taxonomic sense. The vast majority of the former dicots, however, form a monophyletic group called the eudicots or tricolpates. These may be distinguished from all other flowering plants by the structure of their pollen. Other dicotyledons and monocotyledons have monosulcate pollen, or forms derived from it, whereas eudicots have tricolpate pollen, or derived forms, the pollen having three or more pores set in furrows called colpi.

Traditionally the dicots have been called the Dicotyledones (or Dicotyledoneae), at any rank. If treated as a class, as in the Cronquist system, they may be called the Magnoliopsida after the type genus *Magnolia*. In some schemes, the eudicots are treated as a separate class, the Rosopsida (type genus *Rosa*), or as several separate classes. The remaining dicots (palaeodicots) may be kept in a single paraphyletic class, called Magnoliopsida, or further divided. The following lists are of the orders formerly placed in the dicots, giving their new placement in the APG-system and that under the older Cronquist system, which is still in wide use.

APG II	Cronquist system
'Palaeodicots': basal orders	**Magnoliopsida**
Amborellales	Magnoliidae (mostly basal dicots)
Nymphaeales	Magnoliales
Austrobaileyales	Laurales
Chloranthales	Piperales

Ceratophyllales
'Palaeodicots': Magnoliid complex
Magnoliales
Laurales
Piperales
Aristolochiales
Canellales
Eudicots
Basal eudicots
Ranunculales
Buxales
Trochodendrales
Proteales
Gunnerales
Berberidopsidales
Dilleniales
Caryophyllales
Saxifragales
Santalales
Vitales
Rosids
Basal rosids
Crossosomatales
Geraniales
Myrtales
Eurosids I
Zygophyllales
Celastrales
Malpighiales
Oxalidales
Fabales
Rosales
Cucurbitales
Fagales
Eurosids II
Brassicales
Malvales
Sapindales
Asterids
Basal asterids
Cornales
Ericales
Euasterids I
Garryales
Solanales
Gentianales
Lamiales
Unplaced: Boraginaceae

Aristolochiales
Illiciales
Nymphaeales
Ranunculales
Papaverales
Hamamelidae
Trochodendrales
Hamamelidales
Daphniphyllales
Didymelales
Eucommiales
Urticales
Leitneriales
Juglandales
Myricales
Casuarinales
Caryophyllidae
Caryophyllales
Polygonales
Plumbaginales
Dilleniidae
Dilleniales
Theales
Malvales
Lecythidales
Nepenthales
Violales
Salicales
Capparales
Batales
Ericales
Diapensiales
Ebenales
Primulales
Rosidae
Rosales
Fabales
Proteales
Podostemales
Haloragales
Myrtales
Rhizophorales
Cornales
Santalales
Rafflesiales
Celastrales
Euphorbiales
Rhamnales

Euasterids II	Polygalales
Aquifoliales	Sapindales
Apiales	Geraniales
Dipsacales	Apiales
Asterales	Asteridae
	Gentianales
	Solanales
	Lamiales
	Callitrichales
	Plantaginales
	Scrophulariales
	Campanulales
	Rubiales
	Dipsacales
	Calycerales
	Asterales

MONOCOTYLEDON

The Monocotyledons or monocots are a group of flowering plants, dominating great parts of the earth. Monocots comprise the majority of agricultural plants in terms of biomass produced. There are between 50,000 and 60,000 species within this group; according to IUCN there are 59,300 species.The largest family in this group (and in the flowering plants) is the orchids (usually taken to be the family Orchidaceae, but sometimes treated at the rank of order), with about twenty thousand species. These have very complex (and striking) flowers, adapted for highly specific insect pollination.

The economically most important family in this group (and in the flowering plants) are the grasses, family Poaceae (Gramineae). These include all the true grains (rice, wheat, maize, etc), the pasture grasses and the bamboos. This family of the true grasses have evolved in another direction, becoming highly specialized for wind pollination. Grasses produce much smaller flowers, which are gathered in highly visible plumes (inflorescences). A further noteworthy, and economically important, family is the palm family Arecaceae (Palmae).

NAME, CHARACTERS

The name monocotyledons is derived from the traditional botanical name *Monocotyledones*, which derives from the fact that most members of this group have one cotyledon, or embryonic leaf, in their seeds. This as opposed to the traditional Dicotyledones, which typically have two cotyledons. From a diagnostic point of view the number of cotyledons is neither a particularly handy (as they are only present for a very short period in a plant's life), nor totally reliable character. Nevertheless, monocots are a distinctive group. One of the most noticeable traits is that a monocot's flower is trimerous, with the flower parts in threes or in multiples of three. For example, a monocotyledon's flower typically has three, six, or nine petals. Many monocots also have leaves with parallel veins.

Morphology, compared to the (former) dicotyledons

The traditionally listed differences between monocotyledons and dicotyledons are as follows. This is a broad sketch only, not invariably applicable, as there are a number of exceptions. The differences indicated are more true for monocots versus eudicots, as per the APG II system:

Flowers

In monocots, flowers are trimerous (number of flower parts in a whorl in threes) while in dicots the flowers are tetramerous or pentamerous (flower parts are in fours or fives).

Pollen

In monocots, pollen has one furrow or pore while dicots have three.

Seeds

In monocots, the embryo has one cotyledon while the embryo of the dicot has two.

Stems

In monocots, vascular bundles in the stem are scattered, in dicots arranged in a ring.

Roots

In monocots, roots are adventitious, while in dicots they develop from the radicle.

Leaves

In monocots, the major leaf veins are parallel, while in dicots they are reticulate. However, these differences are not hard and fast: some monocots have characteristics more typical of dicots, and vice-versa. This is in part because "dicots" are a paraphyletic group with respect to monocots, and some dicots may be more closely related to monocots than to other dicots. In particular, several early-branching lineages of "dicots" share "monocot" characteristics, suggesting that these are not defining characters of monocots. When monocots are compared to eudicots, the differences are more concrete.

Taxonomy

The monocots are considered to form a monophyletic group arising early in the history of the flowering plants. The earliest fossils presumed to be monocot remains date from the early Cretaceous period.Taxonomists have considerable latitude in naming this group, as the monocots are a group above the rank of family.

Historically, the Monocotyledons Were Named

Monocotyledoneae in the de Candolle system and the Engler system. Monocotyledones in the Bentham and Hooker system and the Wettstein system class Liliopsida in the Takhtajan system and the Cronquist system. Subclass Liliidae in the Dahlgren system and the Thorne system (1992). Clade monocots in the APG system and the APG II system. Each of the systems mentioned above use their own internal taxonomy for the group. The monocotyledons are famous as a group that is extremely stable in its outer borders (it is a well-defined, coherent group), while in its internal taxonomy is extremely unstable .

BASIC FLOWER PARTS

The basic flower parts are broken down into male and female structures that hold the sex cells or gametes. The male flower parts, known collectively as the stamen, consists of the

1. Filament (a type of stem),
2. Anthers (which hold the pollen), and
3. Pollen grains (which deliver the sperm to the female flowers). The female flower parts, known collectively as the pistil consists of the;
    a. Ovary (which contains the ovules which become seeds),
    b. Style (a tube connecting the ovary to the stigma), and
    c. Stigma (the nutrient rich surface where pollen germinates on its way to the ovary).

The Male Journey in Plants

The function of the stamens is to produce healthy pollen with fully formed sperm nuclei. The anthers must be able to open when the pollen is mature, allowing the pollen to be delivered via insect, wind, or simple physical motion of the plant (this is all that is necessary in many "selfers") to the surface of the stigma so that it can enter the pistil. If the pollen is viable and the environmental conditions such as temperature are "right" then the pollen can germinate and form a pollen tube that will grow down the style and deliver a sperm nucleus to an egg cell in one of the ovules in the ovary. The joining of a sperm nucleus with the egg nucleus is the fertilization event that will produce a zygote that becomes the embryo of the developing seed.

The Female Journey in Plants

The function of the pistil is to form an ovary that can have anywhere from one ovule (examples are peaches and avocados) to hundreds of ovules (as in tomatoes or watermelons) in a single ovary. The stigmatic surface of the pistil is where the pollen germinates if the surface is biochemically compatible (*i.e.* the same or closely related species). The growth of the pollen tube that grows down through the style can be a long and arduous journey, with the example of a corn plant's style, its silk, being a particularly striking one. The pollen tube then delivers a pair of sperm nuclei to a single embryo, one to fertilize the egg and the other to fertilize the endosperm. This double fertilization event is very important for all of the true flowering plants (Angiosperms) as the development of the endosperm of the seed is as important to the development of the embryo. All of these events must happen with minimal environmental stress in order for a healthy, vigorous seed to be formed.Floral Biology and Things That Can Go Wrong A number of very routine environmental conditions can occur that will interrupt the seed forming and growing process. Certainly insuring the optimum growing conditions to minimize the environmental stresses is very important during seed development.

Pollen development and production

Seed producing plants must have ample pollen to insure a good seed set. The amount of pollen produced by a particular plant is influenced by two major factors.

Environment

If the plant is stressed due to an inadequate supply of nutrients, water, sunlight, etc., it will at some point depress the development and production of pollen. Also, if the temperatures are too hot or too cold for a particular crop, it can definitely lower or even stop the amount of pollen that is produced.

Genetic

There are definite differences in the amount of pollen produced by different varieties of the same crop. As an example the table beet variety 'Ruby Queen' produces copious quantities of pollen that on a clear day can actually be seen as faint yellow clouds of dust rolling out of a large field of 'Ruby Queen' that's flowering. On the other hand the popular gold coloured beet 'Burpee's Golden Beet' has such poor pollen production that it often sets much less seed than many other table beet varieties.

Pollen Movement

Pollen moves in three ways. In selfers with perfect flowers that remain closed until pollination the plant must be shook for the pollen to release from the anthers and land on the stigma. Usually a little bit of wind in the field will accommodate this, but in the greenhouse it is important to shake the plants each day to release the pollen. In crossers with perfect flowers or monoecious flowers (separate male and female flowers on the same plant) insects are almost always crucial for getting pollen from anthers to stigmas. A shortage of insect pollinizers may then be a serious setback for getting the optimum yields in insect pollinated crossers. Lastly are the crossers that are wind pollinated like beets (with perfect flowers), corn (with monoecious flowers), or spinach (with dioecious flowers, *i.e.* males and female flowers are on separate plants. The wind does all the work blowing the pollen from plant to plant and it is therefore important to have both a good size population of plants so that a lot of potential pollen is present and to have a variety that is a good pollen producer to insure a good seed set.

Pollen Gsrowth

Once the pollen lands on the stigma it must grow down through the style to the ovary. However, if the temperature is either too hot or too cold it may stop the pollen from either germinating on the style or more commonly stop its growth as it goes down the style. Once the pollen tube stops growing it usually doesn't start again and dies before fertilizing an ovule in the ovary. In tomatoes there are some varieties that will set fruit under colder temperatures than others, *i.e.* the Sub-Arctic types, where the pollen is able to grow through cold night temperatures without stopping. Conversely, there are some southern tomato types sold by southern seed companies with names like "Hot Set" where the plants can set fruit even once night time temperatures hit the 90s (F). This phenomenon happens in all species of flowering plants to some degree and can explain why extremes in temperature can often ruin the seed yield of a crop.

Embryo/endosperm Development

Any part of the female flower can suffer under stress. While ovules can be malformed under extremes in environmental stress at any time it is usually during the development of the embryo after fertilization that stress can cause problems that result in poor seed or seed that aborts. Heat, cold, drought or fertility can play a part in this. Heat can be especially detrimental during the rapid growth of the endosperm right after fertilization. Spinach seed is notoriously sensitive to hot weather conditions right as the seed is forming and spinach grown with temperatures much over 80 F usually produce smaller seed and seed that has a lower than usual germination percentage. Many speculate that this is probably due to poor endosperm development under heat stress.

Maturation of Seed

As the seed matures it can be damaged by environmental stresses. Certainly heat and sunscald can damage the germination rate and shelf life of seed. Far more common is the effects of excessive moisture on any of the dry seeded crops (most seeds not borne in fruits) as it can cause bacterial and fungal diseases to spread across the seed bearing portions of the crop that will either affect the quality of the seed outright or will infect the seed with inoculum of the pathogen that can cause disease next year. These seed borne diseases are especially insidious as the grower and seed company may not even know it is there, but when the seed is planted next year the disease may spread through the crop quickly with devastating consequences.

GRASSES

At first sight a grass plant can seem to be rather basic and primitive. However, the grasses are actually quite advanced. There are many different types, each specially adapted to survive in certain conditions. And each has its own particular community of plants and animals associated with it.

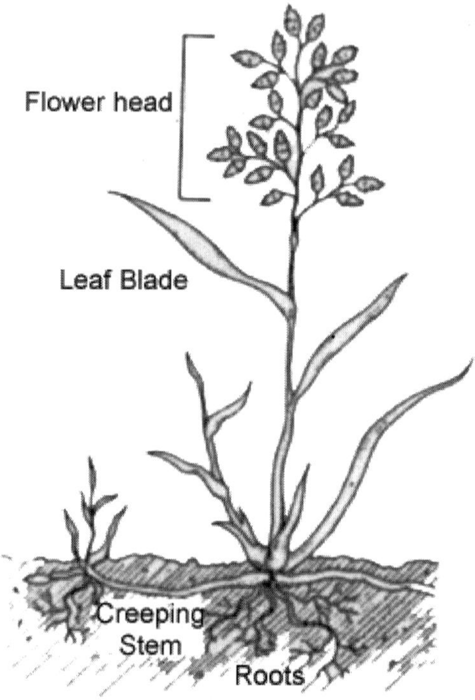

Grass plants can be annual or perennial, and are well adapted to living in difficult conditions. Some can survive droughts, others thrive in wetlands and some withstand extremes of temperature. Annual varieties are very good at colonising land that suddenly becomes available. They grow very rapidly, and within one year produce masses of seed, and then the plants die. Perennial grasses continue growing from year to year. As well as producing seed, they also spread by means of underground stems. In this way the plant can spread across an area, forming a dense mat. This is important in binding soil together and preventing erosion.

Grasses are flowering plants, and belong to the group of plants known as the monocotyledons. These plants only produce one seed-leaf on germination. The plants which we think of as wild flowers belong to the group called dicotyledons, producing two leaves on germination. Grass roots are tough, fibrous and form on clumps or underground mats. Apart from anchoring the plant and absorbing water and minerals, the roots hold the soil together, protecting it from drying out and blowing away, or becoming muddy and washed away.

The stem is cylindrical, hollow and has swollen, solid joints at intervals along it. These are called *nodes*. Each leaf arises from a node on the stem, and then tightly encircles the stem. The way that the stem of a grass grows is part of the reason why grasses are so successful. Some plants form mats, spreading either over the surface of the soil by leafy, creeping stems known as *stolons*, or through the soil by tough scaly stems known as *rhizomes*. One of the main differences between the similar grasses, sedges and rushes lies in the nature of their stems. Rushes usually have pithy stems without solid nodes, and sedges usually have 3-sided, solid stems.

The leaves of different grasses vary in shape and texture, but all grasses have elongated or elliptical leaves which have parallel veins and no stalks. The leaf is really in three parts: the *sheath*, which encloses the stem; the *blade*, or flat, green part; and the *ligule*, where the blade bends away from the leaf. In most plants the main growing point is at the tip of the stem. If this is damaged or cut off, the plant can only grow from buds further down the stem, and this takes some time. In grasses, however, the growing point is near the base of the leaf. If the top of a grass's leaf is cut off, it still keeps growing from the bottom. So when grass is eaten by animals or mown, the grass plants are not killed, but grow even more.

Both monocot and dicot plants are able to reproduce asexually, producing runners, suckers, bulbs and corms which give rise to individuals identical to the parent. This is useful as it allows a plant to spread rapidly, but it is also essential that they have the ability to reproduce sexually. This is how genes are exchanged between plants of the same species, creating genetic variation within a species, the key process behind evolution. Occasionally a variation will be produced that is more successful, and the advantageous gene will be spread through the population. It is this process that enables a species to adapt to changes in the environment. Genetic variation can also help to check the spread of disease, as some members of the species may be disease resistant. One of the problems with genetically uniform crops is that they are very susceptible to disease.

SEXUAL REPRODUCTION

Amongst the flowering plants - grasses and others - there is wide variation in the appearance and form of the flowers. However, the different flowers of these plants all have the same basic structure which allows sexual reproduction to take place. the structure of a generalised flower, with the petals and sepals removed to expose the reproductive parts. The stamen is the male part of the flower and consists of a pollen-producing anther at the end of a fine filament. The female part of the flower is called the carpel. This consists of a stigma, which catches the pollen, connected by the style to the ovary, which contains the eggs. Pollen is deposited on the stigma from another plant, either by wind or by insects visiting the flower. The pollen grain grows a tube down through the style and into the ovary where it enters the egg. The genetic material from the pollen passes down the tube and fuses with that of the egg to create a new individual. The fertilized egg develops into a seed, which can then be dispersed by the plant.

Composite Flowers

There is very wide variation in the form of the flower between the different families. Bur perhaps the most different form of all is found in the composite family. Flowers such as dandelions and daisies are composites, whose flowerheads are made of many small flowers, or *florets*. Each floret has an anther, stigma, style and ovary. Those at the outer edge are known as *ray florets* and have a single long petal attached. The inner florets are known as *disc florets,* and have no petal attached. These features can be observed using a common daisy as an example. Take the daisy apart and look closely with a hand lens to identify the different florets and their parts.

The Structure of Grass Flowers

Grasses are wind-pollinated plants so they do not have brightly coloured or scented flowers to attract insects. Even so, some grass flowers are very beautiful. Grass flowers are in units called *spikelets,* each one containing anything from one to twenty flowers enclosed by two protective bracts called *glumes.* Like other plants each tiny flower has stamens (male parts) and a pistil (female part), but instead of petals there are two small bracts. These often bear bristles, as in the Oat, called *awns*.The whole flower head or *inflorescence* is made up of a number of spikelets.

These are arranged in different ways on the stem according the type of grass. The spikelets may grow directly on the stem, or on stalks. They may be packed close together to form a spike, in tufts or more openly. Plants with showy flowers with strong scents rely on insects to carry the pollen. Since grasses rely on the wind, and much of the pollen will be lost, vast quantities must be produced to increase the chances of success. The stamens are long and the anthers wobbly so the pollen is easily blown off them. The female parts of the flower have long, feathery styles to catch the pollen. As hay fever sufferers know, most grasses flower during May, June and July. However, some grasses flower earlier in the spring, and others continue well into the autumn.

There is even a daily flowering period for grasses. Most grass flowers only open once a day, usually in the morning, some open later in the afternoon, and a very few at midday. This also depends on the weather, as the flowers stay closed on dull or wet days.

ASEXUAL PROPAGATION OF PLANTS

Asexual propagation does not involve exchange of genetic material, so it almost always produces plants that are identical to a single parent. Delicious apple and Bartlett pear are two examples of species that have been asexually propagated for decades. Asexual propagation methods include cuttings, layering, division, grafting, budding and tissue culture.

Cutting

Cuttings involve removing a piece from the parent plant and that piece then regrows the lost parts or tissues. Both woody and herbaceous plants are asexually propagated by cuttings of stems, leaves and roots. New plants can be grown from parts of plants because each living plant cell contains the ability to duplicate all plant parts and functions. Mature cells can change into MERISTEMATIC (mare-ah-ste-MAT-ick) cells that are found at rapid growth sites like buds.There are many types of cuttings. Often, a plant can be propagated by more than one method of cutting. Some plants will reproduce readily from cuttings and others take a considerable amount of time and care. STOCK PLANTS are the parent plants used in asexual propagation. Stock plants must be in excellent health and should possess characteristics desirable for production of new plants. Herbaceous cuttings are those taken from nonwoody plants, such as perennials and houseplants.

Softwood cuttings are pieces of new growth taken from woody stock plants. These cuttings must be taken before the new growth starts to harden. Hardwood cuttings are taken from tissue which has become woody. Other forms of cuttings are leaf cuttings and root cuttings.The gardener must try to duplicate the conditions needed for a plant to root from a cutting. High humidity, indirect light and soil temperatures of 70 to 80 degrees F are best for most cuttings. These conditions may be created by keeping cuttings enclosed under glass or in plastic bags in dappled shade. Cuttings must be shielded from direct sunlight, especially if they are under glass or plastic.

Stem Cuttings

When a cutting is made, injured xylem and phloem cells plug the tubes so that precious fluids are not lost. Usually a callus forms at the cut. Cells near the callus area reorganize to form adventitious roots.Stem cuttings are the most commonly used method to produce houseplants. Select vigorous, new growth with no flower buds. Stem sections should be free of diseases and insects. Each cutting should be 2 to 4 inches long and have 2 or 3 leaves attached.Make a cut 1/4 inch below a leaf node and pull off the leaves that are at the nodes that will be below the surface of the rooting medium. Rooting Hormone helps to stimulate rooting, but is optional. Pour a small amount of the rooting hormone into a clean container to prevent contamination of your entire rooting hormone. Dip the base of the stem, including the node area, into the rooting powder. The stem should be dry when dipped. Commercial rooting products often include a fungicide. This is a good idea given the damp conditions required for rooting success. Tap off excess powder, since too much hormone can inhibit rooting. Poke a hole in the medium before inserting the cutting to avoid loss of the rooting hormone. Insert treated cutting in a moist rooting medium. A suitable rooting medium is half perlite and half sphagnum peat moss. Any disinfested container with drainage is acceptable for use.Cover container and cutting with a plastic bag tent to maintain high humidity. Place unit in a warm area with indirect light. Check the rooting medium every few days to make sure it remains moist. Rooting can take from a few days up to several months.After a few weeks, test for rooting by gently tugging at the cutting. If there is resistance, rooting has started and the plastic cover may be removed.

Leaf Cuttings

In this method, a leaf blade or leaf with petiole is used to propagate new plants. The same steps are followed as for stem cuttings. Choose a healthy leaf from a vigorously growing plant. Cut it close to the stem with a sharp, disinfested razor or knife. Trim off 1/4 of the leaf and dip into rooting hormone, if desired. Insert the leaf into rooting medium so that 1/3 of the leaf is below the surface. One or many new small plants form at the base of the leaf. African violet leaves will produce many new plants. Begonia leaves can be divided into segments for propagation; however each leaf piece must contain a major vein. With leaf cuttings, the original leaf is not a part of the new plant and is usually discarded.

Leaf cutting can propagate many succulent plants, such as sedum, jade, and peperomia. Some plants such as Kalanchoe pinnata or K. bryophyllum, piggyback or air plants, form their own new plantlets in this fashion.

Root Cuttings

Cultivation of root cuttings probably started after gardeners observed new plants growing from pieces of root accidentally left behind in the soil. Take cuttings from newer root growth. Make cuttings 1 to 4 inches long from roots that are 1/4 to 1/2 inch in diameter. Be sure that the roots collected are from the chosen plant and not neighboring plants.Cuttings should be taken during the dormant season when roots have large carbohydrate supplies. However, they also may be taken throughout the growing season. Cut straight through the end of the root closest to the stem. Cut the other end on a slant. This allows you to remember which end is the top (the straight cut) and which is the bottom (the diagonal cut).

Store cuttings from dormant roots for 3 weeks in moist rooting medium at 40 degrees F. Remove from storage and plant upright in the growing medium. Keep moist and warm, in a bright location until growth and weather permit acclimatizing to the outdoors.If root cuttings are taken during active growth, skip the storage period and place cuttings directly in the rooting medium. For smaller plants, take 1- to 2-inch sections. Place cuttings horizontally a half inch below the surface of the rooting medium. These cuttings should be handled indoors or in a HOTBED. The fine roots of many perennials are used for propagation. Root cuttings of some variegated plants will lose their variegation.

Softwood and Hardwood Cuttings

Softwood cuttings are taken from first-year branches that have not yet become woody. Softwood cuttings often propagate flowering shrubs. Late spring and early summer are the best times for success with this method. Take cuttings after it rains or water is applied in the cool morning. Make a diagonal cut. The larger diagonal cut gives more area to develop roots. Keep cuttings in water before moving them into rooting medium.Make cuttings 2 to 10 inches long. Larger cuttings produce larger plants sooner. Make cuts slightly below a leaf node. Follow the same instructions for disinfested containers, rooting medium and light and moisture conditions used by other cutting methods. Rooting hormones may be used with this method. The base of the cutting should be dry before dipping it into rooting hormone powder. Hardwood cuttings are taken once the tissue becomes woody and the plant is dormant. Cuttings can be taken anytime from late fall after a killing frost until late winter. Select healthy wood that was produced the previous summer. Several cuttings can be made from the same branch of some shrubs. Make cuts at a slant, 5 to 12 inches long. Basal cuts should be just below a node, while the upper cut should be slightly above a bud. Dip the basal end in rooting hormone. Mark the most terminal end of each cutting with a tag.

Bury cuttings vertically in moist vermiculite or sand. Cuttings should not freeze, but must remain cool. A callus will form on the lower cut end during storage. Callus formation indicates that cuttings are ready to root.In spring, remove the cuttings from storage. Plant in a hotbed or other protected site with morning sun exposure or filtered light. Leave 1 to 2 inches of cutting above ground. Keep cuttings moist until a root system forms. Transplant the cuttings the following spring while they are still dormant.

Layering

Layering causes roots to develop on shoots that are still attached to the parent plant. The stem is not cut from the main plant until it has rooted. Use this method when other propagation methods are unsuccessful. Layering is a good propagation choice when only a few plants are needed.Simple layering is done by bending a branch to the ground and burying a portion of it. The tip remains uncovered. A light soil increases rooting success, as will wounding or girdling the buried section. Treatment with rooting hormone is helpful.Hold the branch in place with a rock or peg. Layering is done in early spring while plants are still dormant or in late summer on wood that has not become woody.

Plants with flexible branches are particularly suited to this method.Other types of layering include serpentine or compound layering, continuous or trench layering, and mound or stool layering.

Division

Division is the cutting or breaking up of a crown or clump of suckers into segments. Each segment must have a bud and some roots. These segments are replanted and grow into new plants identical to the parent. Most perennials should be lifted and divided when they become overgrown and begin to lose vigor. Vigorous growth in most perennials occurs on the outer segments of the clump. Old growth in the center of the clump is discarded.Carefully dig the plant, loosening the roots and lifting the plant from the soil. Split apart the main clump with two spades or forks or chop with a shovel or hatchet if the clump is firmly massed. In some cases outside segments of the plant can be removed and replanted without disturbing the rest of the plant.A good rule of thumb is to divide fall-flowering perennials in spring and spring- and summer-flowering perennials in fall.Some shrubs that form clumps or crowns are suitable for division. Divide in early spring while dormant. Shrubs used for division should have several clumps that are two years old or older.

Bulbs and Corms

Bulbs can be propagated by removing small bulblets or offsets that form at the base of the parent bulb. These small bulbs take 2 or 3 years to mature into plants that flower. Place offsets in rich, light soil for their development. This same procedure should be followed for plants which form from corms, such as gladiolus. Many lilies can be multiplied by removing scales from the mature bulb. Dust the scale with a fungicide and place, base end down, in a moist growing medium in a warm, protected area. Bulblets will form at the base of the scale. In 1 to 4 years these bulblets will grow and be ready to flower.

Tubers and Rhizomes

Tuberous plants can be dug up and the Tubers separated. In separating the tubers, each must have a segment of the crown that contains at least one eye or bud.Rhizomes grow and develop buds along their length. The rhizomes can be dug and cut into sections that each contain at least one eye or bud.

Grafting

Grafting involves the joining of different segments of two different plants of the same species. A branch or bud is joined to an Understock. In grafting, the cambium layers of the two different segments are aligned and grow together. Some plants graft naturally. Ivy commonly grafts itself. Natural grafting occurs where two branches are in close contact over several years.Grafting allows gardeners to produce plants identical to a parent plant. It also allows growers to control size and shape of a tree or shrub. Apples are grafted onto root stock that regulates the ultimate size of the tree.Grafting can produce more vigorous, earlier fruiting, or two varieties can be grown on the same tree to provide pollination as with apples.On the negative side, some grafting attempts will be rejected. This is called graft incompatibility. This might not occur in the first year, but may in successive years. Wind or winter cold can damage a graft union. Some grafted trees or plants produce large numbers of Suckers.

These can crowd out the desired plant or tree and are unsightly. If the graft union is planted below the soil surface, the scion will root and the rootstock will die. Grafting is usually done in the spring. It involves collecting small branches called scion wood. These branches are 1/4 to 1/2 inch in diameter. Gather them several weeks before grafting is to occur. Collect wood that grew the preceding year. Select only wood with leaf buds, not flower buds. Scion wood should be gathered in winter when wood is dormant, but not frozen. Use sharp clippers so that wood is not crushed. New growth over 1 foot in length is usually best.

Discard the wood at both ends of the branch and use the middle section. End growth is too succulent for grafting and wood closer to the base will graft more slowly. Label the scion wood, wrap it in moist paper towels or sphagnum peat, enclose it in an airtight, plastic container and place it in the refrigerator. Scion wood must be joined to the understock in spring when buds swell. It is critical that the two pieces are nearly the same size and that sap has begun to flow. The day before actually grafting, remove scions from the refrigerator and snip off the bottom ends. Place the clipped scions in a pail of water overnight. It is critical that the cambium layer on the scion precisely matches that of the understock.

The grafted area must be protected from anything that will move the scion out of alignment. Bracing some grafts is advised. Side sprouts should be removed the first summer. Sucker shoots that sprout from below the graft also should be removed. If growth of new graft is satisfactory, do not fertilize the plant during the first year. There is danger that top growth will be very brittle and it could fail to harden off before the first frost.

Budding Or Bud Grafting

Bud grafting is faster, easier and less messy than other forms of grafting. Cambium layers do not need to be aligned. Budwood is collected from healthy branches that grew since spring. Young trees offer good budwood because they produce a large amount of new growth. Use buds from the middle section of the branch. Remove leaves, but keep 1/4 inch of the leaf's petiole for grasping. The bud is cut from the branch and inserted into a T-shaped slice made in the bark of the understock. Budding should only be done when the bark slips easily away from the tree. The bud is held in place with special tape or wrap.

Micropropagation or Tissue Culture

Each plant cell has the potential to grow into a new plant exactly like the parent. This fact coupled with technical advances, specialized equipment and sterile laboratory conditions has produced modern tissue culture. In tissue culture, individual or small groups of plant cells are manipulated so they each produce a new plant. A tiny piece of bud, leaf or stem can produce incredible numbers of new plants in a small space in a short time.The advantages of tissue culture, in addition to speed and efficiency of propagation, include production of disease-free plants. New plants can be made available to the public more quickly because of tissue culture.However, there are some problems with spontaneous mutations which naturally occur.

In tissue culture, the incidence of these mutations is greatly increased.Conditions for tissue culture are very exacting. Absolutely sterile conditions must be maintained. Temperature, light, humidity and atmosphere are strictly controlled with electronic sensors and computerized controls. Such costly equipment rules this out for most home gardeners.

Plant Patents

Many hybridizers now patent their selections. When a plant patent is filed, it is illegal for anyone to propagate the plant for 17 years without written permission from the patent holder.

FLOWERING PLANT REPRODUCTION

Flowering plants have the ability to reproduce both asexually and sexually. While we saw the amount of asexual reproduction decline through the mosses and ferns, many of the flowering plants rely extensively on this means of increasing numbers.

In fact, some flowering plants have lost their ability to reproduce sexually. The asexual reproduction of the flowering plants is discussed in some detail in the laboratory manual. This discussion will be confined to a study of the sexual reproductive cycle. The sexual cycle is probably the most unique and definitely the most successful adaptation of the seed plants. Through their particular cycle, the flowering plants have eliminated the more vulnerable stages found in the life cycles of the mosses and ferns. They are thus able to survive and reproduce in areas where the mosses and ferns cannot survive.

With the immense success of this cycle, the flowering plants have been able to adapt, relatively quickly, to almost any terrestrial environment on Earth. There were three major changes that occurred in the life cycles that produced the success. The first was the development of two different types of spores, or heterospory. The second was the development of the seed. The third was the elimination of the need for water for transfer of gametes. These important features will be discussed at the stages where they occur in the life cycle. To deal with the sexual reproduction and life cycle of the flowering plants we will begin with the dominant or sporophyte generation.

Production of the Sporangia

The reproductive cycle of the flowering plants always begins with the production of the fourth organ or flower. This structure is produced by the sporophyte plants and is the site of production of the sporangia. The flowers consist of a series of modified leaves, similar to the fern fronds. Each is designed to carry out a particular function associated with reproduction. Although some plants may have one or more sets of these modified leaves absent, we will deal with a plant that has a complete flower. The leaves of the outermost whorl are termed the sepals, and the entire group of sepals forms the calyx. The calyx may have several functions but the most common roles are the protection of the developing inner layers and the production of glucose for the future seed development. Sometimes the calyx is coloured to attract pollinators, but usually these leaves are green, reflecting the photosynthetic role of the layer. The next outmost layer of the flower is a whorl of modified leaves termed, petals.

The petals are combined to form the corolla. Again, the corolla may have several functions, all accessory to the actual reproductive cycle, but the most common role is the attraction of pollinators such as birds, bats and insects. It is the corolla that forms the brightly coloured part of the flower.The next layer of modified leaves is the first group of primary reproductive structures. The leaves are called stamens and represent the male part of the flower. The stamens themselves are usually separated into two parts, the stalk or filament and the actual reproductive structures or anthers. It is inside the anthers that the male spores or microspores are produced. Each anther is formed from 1,2 or 4 fused sacs, the microsporangia. During development of the flower, cells of the microsporangia, microsporocytes divide meiotically to produce tiny single celled spores.

These spores in turn divide once mitotically to produce a two celled or two nucleated structure (depending on whether cytokinesis followed mitosis). The structure develops a thick, water resistant coat, and is termed a pollen grain. The haploid pollen grain represents the male gametophyte plant or microgametophyte. At this stage, the pollen grain has a tube nucleus and a generative nucleus, both or neither of which may be surrounded by a cell membrane. The final modified leaf or leaves of the flower are termed carpels. Carpels are the structures that produce the female spores and sporangia. Each carpel or group of carpels develops into a structure called a pistil. Pistils are composed of three parts, the stigma or site of pollen reception, the style and the ovary. The ovary is the structure that produces the female or megasporangium. Inside the ovaries of the pistil are 1 or more ovules. Each ovule consists of layers of protective tissue termed integument and a central cellular core, the megasporangium. At one end, the integument has a tiny, pinhole sized opening, the micropyle. This has a role in insuring successful fertilization. One cell in the megasporangium, the megasporocyte divides meiotically to produce four haploid megaspores. Three of these megaspores disintegrate, but the fourth produces the female or megagametophyte plant.

In review, the sporangia of the flowering plants are therefore located in the flower itself. They are produced by two special types of modified leaves, the stamens and carpels. The two different sporangia, the microsporangia and megasporangia, produce two different types and sizes of spores. The stamens produce many small, male microspores, while the carpels produce few, larger female megaspores. This ability to produce two morphologically and genetically different spores is termed heterospory. The advantage of heterospory is that it permits the development of the different spores in different ways. This greatly increases the developmental options of the plant and means that vulnerable stages can perhaps be eliminated.

The Gametophyte Generation

If we consider what has occurred to this point in the discussion of sexual reproduction, we find that the sporophyte plant has produced sporangia and these in turn have produced haploid meiospores. These cells represented the beginning of the gametophyte generation. At this stage in the mosses and ferns, the spores would have been released to begin an independent development of the haploid gametophyte generation. Here the flowering plants again differ. They have eliminated this vulnerable stage and continue the development of the gametophytes in the protection of the flower of the sporophyte. The gametophyte therefore is never obligated to start an independent existence. Since there are two different spore types, there is a difference in the development of the two gametophytes.

The microgametophyte development begins in the protection of the stamen. It produces the pollen grain and then stops developing until the pollen has been transferred to a receptive stigma through the process of pollination. Once pollination has occurred, chemicals from the female induce further development of the male gametophyte. The pollen grain develops an extension or tube down into the style and ultimately to the ovary and ovules of the pistil. Inside this pollen tube, the generative nucleus divides mitotically to produce two sperm nuclei. These nuclei are the male gametes, and the pollen tube, tube nucleus and sperm nuclei combined, represent the complete and final development of the microgametophyte plant. Note, all this development has occurred in the protection of the flower. The only time the male gametophyte plant was exposed to the environment was during pollination. At that time, the protective outer coat of the pollen grain was intact. The megagametophyte development is quite different, but is similar in the fact that it occurs in the protection of the flower. The megaspore development is induced by the presence of the pollen grain. After pollination, the megaspore develops into the mature gametophyte generation. The single megaspore begins a series of mitotic nuclear divisions to ultimately produce eight nuclei. While this is occurring, the original membrane of the megaspore is enlarging at the expense of the cellular megasporangial tissue. This tissue is enzymatically digested and used by the developing gametophyte. When the eight nuclei are produced, they are positioned at different points in the now enlarged megaspore.

Three of the nuclei are at an end opposite the micropyle. These are termed antipodal nuclei. Each antipodal nucleus develops a cell membrane to become a complete cell. They will play no further role in the reproductive process and are thought to represent the vestigial remnant of the fern prothallus. Three other nuclei are located at the micropyle end of the ovule. These nuclei also develop cell membranes thus becoming three distinct cells. Two smaller synergid cells also play no further role in the development. The remaining larger egg cell is the female gamete and is thus essential for fertilization. Finally, the remaining two nuclei collect in the centre of the megaspore. These polar nuclei do not produce surrounding membranes, but play an important role in flowering plant development.

This concludes the development of the megagametophyte plant. All of this development has occurred inside the flower, but more specifically inside the ovary of the pistil, inside the integument of the ovule, and inside the megasporangium. This protection has provided an excellent environment for ensuring the successful development of the gametophyte plant. The sporophyte plant also, through vascular tissue connections, provides glucose, water and minerals for the developing gametophyte. At this stage the megagametophyte is also termed and embryo sac and consists of 8 nuclei in 7 cells.

Pollination (Gamete transfer)

Pollination is a very distinct and important part of the reproductive cycle. If we compare pollination of the flowering plants to gamete transfer of the mosses and ferns, then the most significant development in the process is the elimination of the necessity for free water for sperm cells to migrate to the egg cells. In a terrestrial environment, the elimination of the requirement for free water when the gametes are mature is extremely advantageous. Flowering plants have instead utilized transfer forces which are more readily available in these environments.

Wind, insects, birds and bats are the major pollinating forces. Each of the flowering plants has evolved a specialized method of enhancing this gamete transfer. For example, plants that rely on wind, produce many light pollen grains and large, feathery stigmas. Plants that utilize insects, have developed bright colours and powerful scents to attract the insects. Whatever the mechanism, the pollination process involves limited exposure of the male gametes and utilizes the existing forces in the environment.

Double Fertilization

Once the pollination has taken place, the megagametophyte matured, and the pollen tube grown to the ovule, the life cycle can continue. The next step is the actual fertilization process. The pollen tube grows from the style to an ovule in the ovary. It is unable to grow through the integument and instead grows around the ovule until it reaches the micropyle. The pollen tube grows through the micropyle and remaining megasporangium to the embryo sac. The point of entry to the embryo sac is right at the position of the egg cell. The pollen tube opens, the tube nucleus disintegrates and the two sperm nuclei are released into the embryo sac. One of the two nuclei is incorporated into the egg cell, fuses with the egg nucleus and produces a diploid, zygotic nucleus and the egg becomes the zygote or first cell of the next sporophyte generation. The other sperm nucleus fuses with the two polar nuclei producing a triploid, endosperm nucleus. Since fertilization is simply described as the fusion of two nuclei, then in fact, in the embryo sac, two fertilization events have taken place. This is termed double fertilization and results in the formation of the 2n zygote and 3n endosperm. At this point several changes in the flower take place. First, the corolla and stamens usually drop off. The stigma and style wither. The ovary wall and sometimes other floral structures begin to develop into the fruit. The fertilized ovule is then termed the seed and the integument develops into the seed coat. The endosperm quickly divides to produce a mass of nutritive endosperm tissue and the zygote divides and differentiates into the embryo.

Development of the Seed

The seed develops directly from the ovule and there are potentially as many seeds capable of being produced as there are fertilized ovules. One of the first events that takes place in seed development, is the rapid growth and division of the endosperm tissue. This triploid tissue take the nutrients provided by the parent sporophyte and stores them. The 3n nature of this tissue permits a more rapid growth than found in the embryo. This tissue will, at some point in the seed development, release its stored food to the embryo. As the endosperm grows rapidly, the embryo begins a slower but more complex development. At first, the zygote divides into a basal and suspensor cell. The basal cell will not develop further but the suspensor cell continues to divide in 1 plane, pushing up into the endosperm tissue, and drawing nutrients from this tissue. After only a few divisions. The terminal cell of the suspensor begins to divide in several planes. This produces a ball of cells called the proembryo. The proembryo continues growth and differentiation into the embryo. The pattern and extent of this development varies among the flowering plants, but in general, the dicots embryo contains an apex which will grow into the plant root (radicle), an opposite apex that will grow into the plant shoot (epicotyl) a transition region between these two apices (hypocotyl) and two masses of storage tissue (cotyledons) or seed leaves. In the dicots, all of the nutrient tissue produced by the endosperm is incorporated into the cotyledons so there is almost no remaining endosperm tissue. In the monocotyledons, there is only one cotyledon produced and a large portion of the stored food reserves of the seed remains in the endosperm. During this time, the seed is growing and the seed coat is enlarging and changing to provide a protective covering for the embryo.

Development of the Fruit

Perhaps the most varied process and structure in the life cycle of the flowering plants is the fruit. The major function of the fruits is to enhance seed dispersal and to protect the seeds. In the terrestrial environment, the seed are the structures of dispersal for the plants. Flowering plants are as adept at using the environment for dispersal as for gamete transfer. Anything that moves in this environment has the potential to be an agent of seed dispersal. Wind, water, and animals are all used by some plants.

Each species has its own modifications for this process and the fruit is usually the most modified and obvious structure involved. For example, plants that use wind as a means of dispersal produce light seed and light feathery fruits with large surface areas to collect the wind. Such plants would include the maple with its characteristic keys and the dandelion with the parachute like tufts. Plants using larger animals as a means of dispersal usually develop one of two strategies. If the seed are to be carried externally, then the fruits develop hooks or barbs for attachment, as in burrs. If the seed are carried internally, then they must be eaten. To ensure this will happen, once the seeds are mature, the fruit ripens into a structure which the animals wish to eat. They are usually brightly coloured, fleshy and sweet tasting. The seeds and sometimes the fruits of these plants must also develop a protective coat which will prevent digestion of the seeds themselves. The seeds will then pass through the digestive tracts of the animal and be dispersed with the faeces.

Seed germination

The final stage in the life cycle of the flowering plant is the germination of the seed into the new sporophyte plant. Again, flowering plants have evolved several mechanisms to better adapt them to the terrestrial environment. First, seed germination does not occur at random, but is rather a timed event to coincide with the most favourable environmental conditions for young plant growth (usually the spring). This time may be cued by temperature, day length or simply time. Only when the proper cue has been received and the is adequate moisture available to support seedling growth, will germination occur. The actual process of germination will be examined in the lab.

MORPHOLOGY AND ANATOMY OF FLOWERING PLANTS

To deal with all the vegetative diversity of the angiosperms would be far more than this course warrants. All of the unique features which allow different plants to exist in some of the more diverse or difficult environments will be omitted. Instead, we will examine some of the more universal features which provide the base of the angiosperm structure. Each plant is made up of four separate organs, three of which are found in all vascular plants, the roots, stems and leaves. The fourth organ, the flower, is unique to this group one further separation of the angiosperms must be made. The flowering plants or Division Magnoliophyta is further divided into class Magnoliopsida (Dicotyledonae) commonly referred to as "Dicots", and class Liliopsida (Monocotyledonae) commonly referred to as "Monocots". The dicots are represented by: the majority of the woody angiosperms such as the oak, apple, elm and cherry trees; the shrubs such as the lilac, spirea and dogwood; and the herbaceous plants such as potato, dandelion, and daisy. The monocots are made up of such plants as the lilies, orchids, bamboo, corn, grasses and reeds. While there are few absolute characteristics to separate these two classes, the majority fit.

COMPARISON OF DICOT AND MONOCOT CHARACTERISTICS

Dicotyledons	Monocotyledons
2 seed leaves (cotyledons)	1 seed leaf
Flower parts in 4's or 5's	flower parts in 3's
Vascular bundle in stem in a ring	vascular bundle scattered
Secondary growth present	secondary growth absent
Taproot system	fibrous root system
Net leaf venation	parallel leaf venation
Leaf usually with petiole	leaf with sheath

GENERAL MORPHOLOGY AND GROWTH PATTERN

While it was easy to describe the morphology of the mosses and ferns, the enormous diversity of the flowering plants makes it almost impossible to describe a 'typical' example. First, these plants can be subdivided morphologically into trees, shrubs and herbaceous plants. The difference in these somewhat arbitrary categories is the general size and amount of secondary or woody growth. Herbaceous plants have little or no secondary growth. They usually remain small and frequently last only one or two seasons.

The perennial plants of this group die-back over the winter or dry periods. Plants such as the cereal grains, potatoes, lettuce, wild flowers, grasses and weeds fall into this category. Shrubs have a much larger portion of their body composed of secondary tissues. These are tissues which develop from lateral rather than apical meristems. The mature tissues are almost exclusively xylem with a minimal amount of phloem. Shrubs are generally larger, and more long-lived than herbaceous plants. Some examples of shrubs are dogwoods, lilacs, chokecherries, raspberries and hazelnut.

Trees represent the largest and longest living of the flowering plants. The bulk of the trees is made up of secondary vascular tissue. Only the growing tips have primary vascular tissue. The xylem, or woody tissue of these plants is produced in concentric annual rings from a band of meristematic cells between the wood and bark. The trees include the maple, oak ash and elm. Bamboo and palms are two examples of monocot trees the herbaceous flowering plants.

These plants illustrate all the successful characteristics of the group and in fact, it is this group that contains the most advanced of the flowering plants. The trees and shrubs are considered to be more ancient organisms and the secondary growth, while it adds bulk to the plants, is not essential to the remarkable adaptability of the group.

Herbaceous plants are separated morphologically into three parts, the roots, stems and leaves. The latter two parts or organs are often lumped together into the shoot. There has been a remarkable evolution of the shoots of these plants, providing special adaptations to almost every conceivable terrestrial environment and to some aquatic habitats. Leaves, for example, have been modified into tendrils for climbing, needles for protection and water conservation, large fleshy areas for water storage and scales for protection. Stems have also been modified as tendrils for climbing, thorns for protection and large fleshy water storage areas.

The roots have undergone the least amount of modification. Roots can be separated into fibrous or tap roots, but they have not been modified for as many different functions as the shoots. Perhaps this is due to the fact the below ground environment is much more stable than above ground.

ANATOMY

The vegetative anatomy of the flowering plants, allowing for some variation, is remarkably consistent form one plant to another. The cell and tissue types are almost identical to the ferns. The flowering plant embryos differentiate early into three types of dividing or meristematic regions. These are outside protoderm, bulk ground meristem and the area that will develop into vascular tissue, the procambium. These embryonic tissues or primary meristems will in turn differentiate into three basic tissue systems. These tissue systems are, outer or dermal system, the vascular tissue system and the remainder of the plant or the ground (fundamental) tissue system. Each of the plant organs has these tissue systems arranged in a particular fashion and comprised of a particular group of cells. The cells and the arrangement depend on the different functions of the organs to the plant as a whole.

The Leaf

The leaf structure is a compromise between three conflicting, environmentally induced, evolutionary pressures. These are: the need to expose the maximum photosynthetic surface to sunlight; the necessity of conserving water; and the need to have an adequate gas exchange between the inside and outside leaf environment to permit photosynthesis. If we deal only with regular photosynthetic leaves. They appear to be separated into three distinct regions, the epidermis, the mesophyll and the veins or vascular bundles.The epidermis is similar to that of the ferns. The tissue that makes it up is epidermal and the cell type is also termed epidermal. The same modifications of cuticle and guard cells exist for the same reasons. The mesophyll, made of fundamental tissue and chlorenchyma cells, is usually more frequently organized into distinct palisade and spongy layers than is found in the ferns.

The vascular bundles are similar to those of the ferns, but three new cell types appear in the vascular tissue of the flowering plants. In the ferns, the xylem was made of conducting elements called tracheids, while in the flowering plants there is also another type of conducting unit termed the vessel. Vessels begin their development in a similar fashion to the tracheids, but at maturity, they lose their end walls. This forms a continuous tube so the water and minerals do not have to travel in a zig-zag fashion but rather can travel straight through the leaf. These tubes of vessels are continuous with the vessels in the stems and roots and provide a more rapid and efficient route of transport.

The phloem of the flowering plants is organized in a similar fashion to the vessels of the xylem. Instead of sieve cells, the phloem contains sieve tube elements or cells. The difference between these and the sieve cells is again because of the perforation of the end walls and not a solid end plate with sieve units on the lateral walls. The sieve tubes are anucleate at maturity but they still contain a living functional cytoplasm. The control of the activity of this cytoplasm is brought about by a group of connected adjacent cells termed companion cells. These tiny, elongate cells direct the sieve tube activity through plasmodesmatal connections. The flowering plants do not have the sieve cells. Vascular bundles of the leaves are usually surrounded by a special, single cell layer termed a bundle sheath. The parenchyma cells of this sheath are responsible for transferring large amounts of glucose that has been produced in the leaf mesophyll, to the phloem.

The Stem

In many respects, the stems of the flowering plants are similar to the stems of ferns. The variations come partially from the need to adapt to above ground conditions and partially from the need to adapt to the wider range of habitats occupied by the seed plants. There are two main patterns of organization in the stem anatomy. The dicotyledons present one anatomical pattern while the other is found in the monocotyledons. The monocotyledons stems have only three regions, an epidermis, cortex and vascular bundles scattered throughout the cortex. The dicots have four regions. The epidermis, cortex, a ring of vascular bundles and a central pith.

The epidermis is different from that of the ferns since it now must protect the stem from the hazards of an above ground environment. Also, the stems of these plant are frequently photosynthetic so guard cells are also required.

The cortex varies according to the plants and the environment. Plants in drier regions often have many sclerenchyma cells in their cortex, while those of more moderate areas have parenchyme (chlorenchyma) and collenchyma. Collenchyma cells are living, strengthening cells which have increased cellulose fibres in the walls. These cells are usually found at the outer surface of the stems, frequently in the corners. The pattern of organization of the vascular bundles of the dicots is a ring.

The vascular tissue is again subdivided into xylem and phloem and these two subtissues have the same cell types as in the leaves of the flowering plants. The xylem of the vascular bundles always is located on the inside of the phloem, but the phloem never rings the xylem as in the fern rhizome vascular bundles. The scattered bundles of the vascular tissue of the monocots has proportionately more vessels than that of the dicots.The pith, found only in the dicots, comprises the central region of the stem and is made up of parenchyma cells. These cells, rich in leucoplasts, are adapted to store the products of photosynthesis and water.

The Roots

Roots have the least cell and tissue differentiation of all the organs. They function in absorption, storage, transport and anchoring. The absorption properties are found only in the extreme tips of the roots and root branches. After approximately 10 - 20mm from the tip, the root tissues mature and water and minerals can no longer be absorbed. The root epidermis develops a cuticle to prevent loss of the water already absorbed and to reduce the affect of harmful soil organisms.

At the extreme tip of the roots is a layer or cap of constantly replaced dead cells. This root cap protects the root as it grows through the soils in search of sufficient water and minerals. Behind the root cap is a zone where water is absorbed. To facilitate this process, many of the epidermal cells in this region develop projections out from their outer walls termed root hairs. The root hairs are lost after the area of absorbtion in replaced by more mature tissues. The outer region of the roots is again the epidermis.The cortex occupies the bulk of the root.

It is often quite well developed with large numbers of parenchyma cells. These cells contain many leucoplasts and are the site of starch storage. Biennial plants store the nutrients for over winter in this region. The inner layer of the cortex is the Casparian strip. This layer functions in a similar fashion to the comparable layer of the fern rhizome.

The vascular bundle or stele of the root is located in a central position. The outer cell layer, termed the pericycle is formed from relatively undifferentiated parenchyma cells. These cells can easily revert back to meristematic or dividing cells. If a root branch is required, it originates from this layer. This allows the newly forming vascular tissue of the branch to link up with the vascular tissue of the primary root and eventually form part of the continuum of the vascular system that runs throughout the entire plant. The xylem is again central to the phloem, as it was in the stem. The cell and tissue types are the same.

11

Germination Test Environments

As mentioned above, germination test prescriptions are available for only a small proportion of seed-producing species. What environment should be chosen to test the germination of seed accessions of species for which no prescription or no acceptable prescription is available? This is exactly the problem which faced staff at the Royal Botanic Gardens Kew, Wakehurst Place Gene Bank, for almost all their accessions, because this gene bank deals entirely with seed collections of species which do not have great commercial significance. The method that they developed to answer the question posed above provides another approach to developing appropriate germination test regimes for seed accessions maintained in gene banks.

SUITABLE TEST ENVIRONMENT

This approach acknowledges that the most suitable environment for testing each accession is unknown. Thus, instead of a single prescription, an attempt has been made to develop the shortest series of tests which is likely to give a solution to the problem of defining a suitable test environment. The decision as to which test to try next depends on the results obtained in the previous test(s), and so we call this series of tests an algorithm. For gene banks a truly universal algorithm would be one appropriate to all seed-producing species, irrespective of taxonomy. However, such an approach would be rather unwieldy and require a large number of tests. Certain similarities in the response of seed germination to environment for species within a family can be recognised. Consequently this approach provides a separate algorithm for each family where sufficient knowledge has been accumulated.

Where available, the algorithms developed by the staff at the Wakehurst Place Gene Bank are presented according to family. However, it is necessary here to provide some explanation of how the algorithms were developed, their possible faults, and how they should be applied.

THE DEVELOPMENT OF ALGORITHMS

Storage at the Wakehurst Place Gene Bank began in 1974. Since then all accessions entered into storage have been tested for germination as soon as possible thereafter. In many cases accessions were tested in more than one germination test environment, the reason for this being an inadequate proportion (less than 85 per cent) of seeds germinating in the first test.

Thus for many accessions it was possible to compare germination test results over a number of different environments with and without the imposition of various additional dormancy-breaking treatments. For each accession the best germination test regime was noted, and this regime was then used for all subsequent germination tests of that accession.

In addition the results of all the different germination tests carried out on each accession were filed according to species to provide a growing information bank on the response of seed germination to different test conditions. The below table provides an example of the information generated and filed from such tests for a single accession of Veronica verna.

Table Example of germination test results recorded by staff at the Wakehurst Place Gene Bank used in subsequent analyses to determine algorithms for determining suitable test regimes: germination (per cent) and classification of non-germinating seeds after testing seeds of a 1977 collection of Veronica verna under four different regimes (100 seeds per test).

Test Date	Germination Test Regime	Total Test Duration (Days)	Cumulative Germination (%) at End of Test	Non-germinated Seeds at End of Test[1]		
				Fresh	Mouldy	Empty
16/12/77	21°C	42	0	100	0	0
16/12/77	21°/11°C (12h/12h)	157	48	52	0	0
18/4/78	26°C for 28 days (imbibed) then 11°C	48	100	0	0	0
18/4/78	2°C	73	88	12	0	0

Note:

[1]expressed as the percentage of the original number of seeds tested

The two original tests on seeds from this accession, begun on 16/12/77, demonstrated a high proportion of dormant seeds (those designated as fresh at the end of the test. At 21°C all seeds exhibited dormancy (Table). The alternating temperature 'regime of 21°/11°C was only partly successful in breaking this dormancy.

Therefore towards the end of the latter test, a decision was made to sample more seeds from the accession and begin two additional germination tests in different conditions (which are known to promote germina-tion in several species).

One of these additional regimes-consisting of a preliminary period during which the imbibed seeds were exposed to 26°C (sometimes described as a warm stratification treatment) followed by subsequent exposure to 11°C-was entirely successful at breaking dormancy, enabling all the seeds tested to germinate. Consequently this regime will be used for any subsequent germination tests to monitor the viability of this accession.

Over the past ten years or so a considerable amount of information on the response of seed germination to various test regimes (similar to that described above for Veronica verna) has been generated by staff at Wakehurst Place. This raw information has subsequently been analysed, principally by Mr. S. Linington, in the following manner. For a particular family all the routine test data for all accessions (each similar in form to that provided in the above Table) were examined. It was (arbitrarily) decided that test regimes in which 85 per cent, or more, of the full seeds germinated would be classified as successful. For each accession all the successful regime(s) (if there were any) were listed. For example, two regimes would be listed as successful for the accession shown in Table: 2°C constant; and 28 days at 26°C followed by 11°C constant.

When this first stage of analysis was completed a pattern of successful regimes within each family was sought. Emphasis was placed on determining those simpler regimes (that is those employing few dormancy-breaking agents) which were successful for the greatest number of accessions. These regimes are those which are now used as the first step in the algorithms which have been developed. More complicated regimes (involving the use of many dormancy-breaking treatments) which are successful for fewer accessions are then provided as subsequent alternatives (later steps) in the algorithm if the first step does not provide a suitable environment for germination testing.

LIMITATIONS OF THE ALGORITHMS

There are a number of possible faults inherent in this method of analysis. First, by excluding from the analysis all accessions which failed to reach 85 per cent germination under any test regime, the response of accessions which are either very dormant, or very poor quality and vulnerable in certain germination test environments may have been ignored. Secondly, it is based on tests of accessions within the Wakehurst bank which, at least in the past, were biased towards collecting in temperate regions. Not only may this result in an algorithm more suited to temperate collections, but it may also result in algorithms more suited to certain tribes within each family. For example, the algorithm for the Leguminosae was based largely on accessions of the Papilionoideae tribe and little data was available for accessions from either the Caesalpinoideae or Mimosoideae tribes. Finally it should be noted that the analysis was based on data obtained over only a limited range of treatments. Other, untested, treatments may be superior and there is as yet no way of knowing whether this will be the case.

SUCCESS OF THE ALGORITHMS MAY VARY BETWEEN FAMILIES

Despite these reservations, the algorithm approach to determining germination test environment works well at Wakehurst and is likely to be useful in other gene banks. Tests at Wakehurst on previously untested accessions using the algorithms for the Amaranthaceae, Gramineae, Leguminosae and Solanaceae families have shown that 100 per cent, 50 per cent, 90 per cent and 75 per cent of accessions within each family reached 85 per cent germination or greater.

In view of the importance of the Gramineae in food production the limited success of the algorithm for this family is disappointing. One group of species within the Gramineae whose seed is particularly difficult to germinate is the tropical pasture species. Unfortunately tests on five seed lots of tropical pasture species at Reading confirm that the present algorithm for the Gramineae is not particularly helpful for determining germination test environment in these species.

DEMONSTRATION OF THE USE OF AN ALGORITHM

Nevertheless, Table can be used to explain the method of applying the algorithm. Table should be studied together with the Gramineae algorithm provided in Chapter. In Table every feasible combination of dormancy-breaking agents within the algorithm was applied to the seeds, amounting to some 24 separate test environments. Note, however, that only a few of these tests would have been applied if the algorithm had been followed. The pathway through the tests which would have been followed if the Gramineae algorithm had been applied is shown by the boxed values in Table, the sequence being from the top of the table to the bottom. Table Normal germination (expressed as percentage of full seeds tested) after 28 days in germination test [lower figures in square brackets after 56 days] for five tropical grass seed lots. If the Gramineae algorithm had been followed in the manner intended only the tests in square boxes would have been investigated.

SEED DORMANCY AND MORPHOLOGY

The fruits are usually surrounded by a persistent perianth. B.R. Atwater classifies seed morphology as endospermic seeds with peripheral linear embryos. Light, alternating temperatures, pre-washing and removal of the seed covering structures can promote germination, but some care may be required in defining promotory treatments in detail. Detailed information on seed germination is provided for the genera Amaranthus and Celosia in this chapter. A limited number of other recommended germination test procedures is provided in Table and the algorithm below may be helpful in developing suitable germination test procedures.

RBG KEW WAKEHURST PLACE ALGORITHM

The first step of the algorithm is to test three samples of the seeds at constant temperatures of 26°C, 31°C, and 36°C with light applied for 12h/d. If full germination has not been promoted and there is a trend in germination response to constant temperatures, then test further samples of seeds at more extreme constant temperatures.

That is: if germination was greatest at 26°C then test at constant temperatures of 16°C and 21°C with light applied for 12h/d; but if germination was greatest at 36°C then test at a constant temperature of 41°C with light applied for 12h/d. If, however, germination was greatest at 31°C then go to the next step of the algorithm.

If full germination has not been promoted, the second step of the algorithm is to test a further sample of seeds in an alternating temperature regime of 33°/19°C (12h/12h) with light applied for 12h/d during the period spent at the upper temperature. If full germination has not been promoted, the third step of the algorithm is to estimate viability using a tetrazolium test.

If the result of the tetrazolium test indicates that the failure to achieve full germination is due to the presence of dead seeds and that one of the above regimes promoted the germination of all, or almost all, the viable seeds, then this regime is used for all subsequent germination tests. If, however, the result of the tetrazolium test indicates that dormancy has not been broken by the regimes applied so far in the algorithm, then experiment with modifications to the above regimes.

Evidence of Dormancy

Almost without exception seeds of Amaranthus spp. show considerable dormancy. This is possibly because the majority of references cited here concern seeds from weedy plants of Amaranthus spp. Only in a minority of papers has the dormancy of seeds from cultivated plants within the genus Amaranthus been investigated. Nevertheless it is clear that dormancy can also be considerable in seeds from cultivated plants-*e.g.*

After-ripening treatment periods required to remove dormancy from the majority of seeds can be considerable: 180 days (4,22), 300 days or as many as 4 years for seeds of A. retroflexus; and 6 months for seeds of A. blitoides and A. graecizans.

Comment

Comparatively few treatments have been reported which are able to break the dormancy of all seeds within accessions ofAmaranthus spp. Light is a factor of critical importance. At first sight the literature may appear confusing, since the effect of light on germination has been reported as both stimulatory and inhibitory. Inhibitory action occurs when the light is of the wrong wavelength-for example, far red-or the dosage is too high. To obtain the maximum promotion from light additional parameters are important: these include tempera-ture, the timing of the light application, and-more occasio-nally-the presence of other dormancy-breaking agents.

At constant temperatures, high temperatures are required for the germination of dormant seeds: above 30°C, 30°–37°C, 35°C, 35°–40°C, or 40°C. Alternating temperatures can promote germination: 20°/35°C (16h/8h) is reported to be superior to 20°/30°C (16h/8h)-the AOSA prescribed regime-whilst both are better than high constant temperatures; 30°/25°C (day/night) is superior to 25°/20°C, 20°/15°C, or 15°/30°C (day/night); 25°/8°C (18h/6h) is reported to be better than constant temperatures between 25° and 28°C; and 23°/35°C (12h/12h) is superior to constant temperatures of 30°C or 35°C, but to obtain maximum promotion the initial imbibition temperature should be 23°C, not 35°C-that is 23°/35°C, not 35°/23°C.

However, neither constant nor alternating temperatures alone can be regarded as satisfactory. For example, testing dormant seeds from five Amaranthus spp. at constant temperatures between 29°C and 38°C, or alternating temperatures of 19°/23°C, 20°/32°C or 22°/43°C in the dark resulted in no more than 10 per cent germination in any one environment, with no significant difference in germination between the various germination test regimes.

At high constant temperatures the germination of dormant seeds is promoted by only a very short duration light treatment, for example in A. blitoides and A. retroflexus. As little as 45–60 seconds can be sufficient to promote the germination of the majority of dormant seeds-for example, promoting germination from 25 per cent to 92 per cent.

However, the timing of the light treatment in relation to the start of imbibition is critical: in A. hybridus the maximum promotion is reported after 80-150 hours; in A. blitoides after 48 hours; and in A. retroflexus after 96 hours. Moreover, the response to light can be considerably increased in the presence of other dormancy-breaking agents-particularly potassium nitrate.

Although the above times appear to vary—probably due to differences of irradiance reaching the seeds-once the maximum sensitivity to light has been achieved during imbibition, delaying the light treatment does not reduce the germination of seeds of Amaranthus spp.

Given the wide differences in the level of seed dormancy between accessions within the genus Amaranthus it is difficult to make precise recommendations. Nevertheless, the following regime is tentatively suggested: imbibe seeds in the dark at 35°C for four days with 0.2 per cent potassium nitrate co-applied; expose briefly to red light-1×10^{-9} mol cm^{-2} is probably sufficient.

Although one exposure is probably satisfactory, it might be worthwhile to provide this daily in case some individual seeds require a longer period of imbibition before they reach maximum sensitivity to light.

It is likely that an alternating temperature regime would be preferable to a constant temperature of 35°C; 20°–25°/35°C (12h/12h) is suggested, again with potassium nitrate and red light as described above, but this must be regarded as untested as yet. Finally, if it is difficult to provide a suitable light environment it is worth noting that pre-chill treatments can reduce the need for light in subsequent germination tests; try 14 days at 3°–5°C and then transfer to 35°C in the dark for the germination test.

GERMINATION TO THE TEMPERATURE REGIME

In addition to all the possible germination test regimes according to the Gramineae algorithm, Table also presents results of tests carried out in two additional alternating temperature environments (15°/30°C and 10°/30°C) and also in combination with 10^{-3} M KNO_3 (where sufficient seeds were available). These results suggest that, for some seed lots at least, those gene banks testing tropical grass species would be well advised to include additional alternating temperature regimes as the second step of the algorithm. For three of the seed lots tested here small differences in the alternating temperature regime greatly influenced the proportion of seeds which germinated. Not only did this result in almost full germination of viable seeds, but it would have avoided the need for several stages of the algorithm in the case of Gatton Panic.

In passing note that the results shown in Tables emphasise that seed germination can be particularly sensitive to the temperature regime in which the seeds are tested. It cannot be overemphasised that gene bank staff must make strenuous efforts to get this right first before embarking on investigations as to the efficacy of other dormancy-breaking agents.

POSSIBLE USE OF UNGERMINATED SEEDS FROM PRIOR STEP IN ALGORITHM FOR THE CURRENT TEST

In the algorithms provided here additional tests are carried out by drawing a fresh sample of seeds from the accession bulk in storage. In some cases, for example if the supply of seeds is severely limited, it is possible to test the response to additional treatments by applying these to those ungerminated seeds which remain fresh and firm at the end of a previous test.

This is particularly suitable for scarification, dehusking or puncturing treatments since it may avoid the need to apply these time-consuming procedures to all seeds tested. For example, if 70 per cent of seeds tested of a legume accession germinate without scarification it is easier to scarify the remaining proportion of seeds and to return these to the germination environment than to draw a fresh sample of seeds from storage and scarify all of the seeds to be tested.

DURATION OF GERMINATION TESTS

During each step of the algorithm it is necessary to decide when to terminate the germination test and whether to pursue the next step in the algorithm. Since both aged and dormant seeds may only begin to germinate in a test after a substantial delay it is not possible to provide categorical advice. In Chapter 10, Volume I, an example is provided of how to decide when to terminate a germination test by constructing a germination progress curve.

This technique can be used for each step in the algorithm, but each test should not be continued for an excessive period because the next step in the algorithm might result in a more suitable and more rapid procedure. That is if the progress of germination is very slow it might be better to pursue the next step in the algorithm.

A minimum germination test period of 28 days is suggested. A comparison of results after 28 and 56 days in test, for example, shows substantial germination after 28 days in test for only the Brachiaria humidicola accession tested at 35°/20°C. Consequently it is suggested that the decision to test a further sample of seeds in the next step of the algorithm can generally be made once the 28-day test result is known, but that the current test be continued for a further period dependent upon the shape of the germination progress curve. In tests at a very low temperature (*e.g.* 6°C), however, it may be better to delay a decision as to whether to continue through the algorithm until the 56-day test result is known.

USE OF ALGORITHMS AT ACCESSION RECEIPT

It is suggested that the algorithms could be applied soon after an accession is received if the standard advice on germination test procedures and dormancy-breaking treatments is unsatisfactory or if no advice is available. The number of seeds tested in each of the regimes required by the algorithm need not exceed 100 (and 50 seeds may well be sufficient), but the seeds should be sampled from the accession at random and replication in each test is advisable Once a suitable test procedure has been developed a further sample of 200 seeds can be withdrawn from the accession and tested by this procedure in order to 'estimate initial accession viability. Subsequent monitoring tests of accession viability will also apply the same germination test procedure, but may be either fixed sample size tests or sequential tests. Note that it is envisaged that the algorithms would only be applied once, *i.e.* when the accession is first received: the subsequent monitoring tests would employ the most appropriate regime already determined.

SEED DORMANCY AND THE ROLES OF ENVIRONMENTAL FACTORS

During seed maturation, the embryo enters a quiescent phase in response to desiccation. Seed germination can be defined as the resumption of growth of the embryo of the mature seed; it depends on the same environmental conditions as vegetative growth does. Water and oxygen must be available, the temperature must be suitable, and there must be no inhibitory substances present.

However, in many cases a viable (living) seed will not germinate even though all the necessary environmental conditions for growth are satisfied. This phenomenon is termed seed dormancy. Seed dormancy introduces a temporal delay in the germination process that provides additional time for seed dispersal over greater geographical distances. It also maximizes seedling survival by preventing germination under unfavourable conditions.

Two types of seed dormancy have been recognized, coat-imposed dormancy and embryo dormancy. Coat-imposed dormancy is dormancy imposed on the embryo by the seed coat and other enclosing tissues, such as endosperm, pericarp, or extrafloral organs. The embryos of such seeds will germinate readily in the presence of water and oxygen once the seed coat and other surrounding tissues are either removed or damaged.

There are five basic mechanisms of coat-imposed dormancy:

- *Prevention of water uptake.* Prevention of water uptake by the seed coat is a common cause of seed dormancy in families found in arid and semiarid regions, especially among legumes, such as clover (*Trifolium* spp.) and alfalfa (*Medicago* spp.). Waxy cuticles, suberized layers, and lignified sclereids all combine to restrict the penetration of water into the seed.
- *Mechanical constraint.* The first visible sign of germination is typically the radicle breaking through the seed coat. In some cases, however, the seed coat may be too rigid for the radicle to penetrate. Nuts with hard, lignified shells are examples of dormancy caused by mechanical constraint. Such shells must be broken by biotic or environmental forces for the seed to germinate. Even nonlignified tissues, such as the endosperm of lettuce seeds, can suppress expansion of the embryo. For the seeds to germinate, the endosperm cell walls must be weakened by the production of cell wall–degrading enzymes.
- *Interference with gas exchange.* Some seed coats are considerably less permeable to oxygen than an equivalent thickness of water is— *e.g.*, less by a factor of about 10^4 in seeds of cocklebur (*Xanthium pennsylvanicum*). This lowered permeability to oxygen suggests that the seed coat inhibits germination by limiting the oxygen supply to the embryo. In support of this idea, investigators can break the dormancy of such seeds either by making a small hole in the coat with a pin (without weakening the coat mechanically), or by treating the coat with concentrated oxygen. However, other studies suggest that the oxygen consumption of the embryos from such seeds is considerably less than the amount of oxygen able to penetrate the seed coats under normal aerobic conditions. Thus the role of oxygen impermeability in seed coat dormancy remains unresolved.
- *Retention of inhibitors.* The seed coat may prevent the escape of inhibitors from the seed. For example, growth inhibitors readily diffuse out of isolated *Xanthium* embryos, but not from intact seeds.
- *Inhibitor production.* Seed coats and pericarps may contain relatively high concentrations of growth inhibitors that can suppress germination of the embryo. ABA is a common germination inhibitor present in these maternal tissues. In certain cases where repeated washing (leaching) removes dormancy, the effect is thought to be due to the loss of such inhibitory compounds.

The second type of seed dormancy is embryo dormancy, which refers to a dormancy that is inherent in the embryo and is not due to any influence of the seed coat or other surrounding tissues. In some cases, embryo dormancy can be relieved by amputation of the cotyledons. Species in which the cotyledons exert an inhibitory effect include European hazel (*Corylus avellana*) and European ash (*Fraxinus excelsior*). A fascinating demonstration of the cotyledon's ability to inhibit growth is found in species (*e.g.*, peach) in which the isolated dormant embryos germinate but grow extremely slowly to form a dwarf plant. If the cotyledons are removed at an early stage of development, however, the plant abruptly shifts to normal growth.

Embryo dormancy is thought to be due to the presence of inhibitors, especially ABA, as well as the absence of growth promoters, such as GA (gibberellic acid). The loss of embryo dormancy is often associated with a sharp drop in the ratio of ABA to GA.

SEED DORMANCY MAY BE PRIMARY OR SECONDARY

Different types of seed dormancy also can be distinguished on the basis of the timing of dormancy onset rather than the cause of the dormancy:

- Seeds that are released from the plant in a dormant state are said to exhibit primary dormancy.
- Seeds that are released from the plant in a nondormant state but which become dormant if the conditions for germination are unfavourable exhibit secondary dormancy. For example, seeds of *Avena sativa* (oat) can become dormant in the presence of temperatures higher than the maximum for germination, whereas seeds of *Phacelia dubia* (small-flower scorpionweed) become dormant at temperatures below the minimum for germination. The mechanisms of secondary dormancy are poorly understood.

ENVIRONMENTAL FACTORS CONTROL

Various external factors release the seed from dormancy, and dormant seeds typically respond to more than one condition. Many seeds lose their dormancy when their moisture content is reduced to a certain level by drying. This method of breaking seed dormancy is called afterripening, and is usually performed in a special drying oven. On the other hand, if the seed becomes too dry (5 per cent water content or less), the effectiveness of afterripening is diminished. Another factor that can release seeds from dormancy is low temperature, or chilling.

Many seeds require a period of cold (0 to 10°C) while in a fully hydrated (imbibed) state in order to germinate. In temperate-zone species, this requirement is of obvious survival value, since such seeds will germinate not in the fall, but only in the following spring. Chilling seeds to break their dormancy is a time-honoured practice in horticulture and forestry and traditionally has been referred to as *stratification*. This term is derived from the old agricultural practice of allowing seeds with a chilling requirement to overwinter outdoors in layered mounds of moist soil. Today the seeds are simply stored in a refrigerator.

The third external factor that plays an important role in breaking seed dormancy is light. Many seeds have a light requirement for germination, which may involve only a brief exposure, as in the case of lettuce, an intermittent treatment (*e.g.*, succulents of the genus *Kalanchoe*), or even a specific photoperiod involving short or long days.

As was discussed, phytochrome is the main sensor for light-regulated seed germination. Interestingly, all light-requiring seeds exhibit seed coat dormancy, and removal of the outer tissues of the seed allows the embryo to germinate in the absence of light. The effect that light has on the embryo is thus to enable the radicle to penetrate the seed coat. This penetration often involves some enzymatic weakening of the enclosing tissues.

The Longevity of Seeds

How long can a seed remain dormant and still remain viable? Seed longevity is of practical importance because of ongoing efforts to preserve plant genetic resources for future agricultural crops by setting up seed gene banks. Spectacular claims of dormant seeds that remained viable for thousands of years have been made, but are considered highly controversial. Extreme examples of ancient seeds that have been reported to retain their viability include submerged lotus seeds found in a 3,000-year-old boat near Tokyo, barley seeds from the 3,000-year-old tomb of King Tutankhamen, and arctic lupine seeds associated with rodent burrows determined to be 14,000 years old. In most cases, however, a scientifically rigourous examination of the data has either disproved or raised serious doubts about the claimed antiquity of the seeds.

The most serious criticism is that the dates of the seeds have been inferred from the age of their immediate surroundings or associated artifacts rather than from direct measurements of the seeds themselves. This question raises the possibility that the seeds are much younger than the site at which they were found. In fact, microscopic examination of authenticated ancient cereal grains from various Middle Eastern sites has shown that cells and tissues of these seeds are no longer intact. Such seeds are clearly inviable.

Although the most sensational claims for seed longevity are almost certainly bogus, seeds of *Canna compacta* apparently can live for at least 600 years. Viable *Canna* seeds were obtained from inside a walnut in a tomb in Argentina. The *Canna* seeds had apparently been inserted into the immature seeds of a growing walnut fruit before the hard outer shell formed. Once the shell hardened and the nut dried out, the result was a rattle. Native people strung the rattles together to form a necklace. In this case, the seeds had to be at least as old as the walnut shell, and carbon dating of the shell indicated that it was about 600 years old.

Plant collections begun during the late eighteenth century in Europe—for example, at the British Museum in London and at the Museum of Natural History in Paris—have been reliable sources of seeds for viability determinations. *Cassia multijuga* (false sicklepod) seeds, collected, were still viable after being tested, W. J. Beal initiated the longest-running experiment on seed longevity by burying the seeds of 21 different species in unstoppered bottles in a sandy hilltop near the Michigan Agricultural College in East Lansing. After 100 years, only one species, *Verbascum blattaria* (moth mullein), remained viable.

Why are some seeds able to survive in a dormant state for hundreds of years, whereas others lose viability in less than 5 years? Almost nothing is known about the mechanisms that determine the longevity of seeds. If these mechanisms could be understood, we might someday be able to greatly increase the seed longevity of agriculturally important species and varieties, thereby enhancing our ability to preserve plant genetic resources for generations to come.

EMBRYO DORMANCY

Embryo dormancy (embryo component of non-deep PD): A dormant embryo is characterized by a high ABA/GA ratio, high ABA sensitivity and low GA sensitivity. Embryo dormancy release involves remodelling of hormone biosynthesis and degradation towards a low ABA/GA ratio, a decrease in ABA sensitivity and an increase in GA sensitivity. Thus ABA dominates the embryo dormancy programme, and GA the embryo germination programme. A non-dormant embryo is characterized by increased growth potential, the ability for cell extension growth and the ability to induce the release of coat dormancy.

COAT DORMANCY

Coat dormancy (coat component of non-deep PD): The combination of an embryo with low growth potential and mechanical constraint from the seed covering layers can result in dormancy. GA can release this coat dormancy by increasing the embryo growth potential and/or by reducing the mechanical constraint.

There are two forms which are distinguished based on whether the covering layer consists of dead (*e.g.* outer testa) or living (*e.g.* endosperm) tissue:

- *Testa dormancy*: ABA during seed development determines the subsequent GA requirement for germination since ABA influences the testa characteristics (*e.g.* thickness) and GA the embryo growth potential.
- *Endosperm dormancy*: Endosperm weakening can be either part of the coat dormancy release or part of the germination programme. Since

the endosperm is in most cases living tissue it can actively participate in regulating embryo constraint by influencing both the ABA/GA ratio and sensitivity to these hormones. GA acts by increasing the embryo growth potential and by promoting endosperm weakening which is achieved though ABA-independent and ABA-inhibited mechanisms.

Induction, Maintenance and Release of Physiological Dormancy by Plant Hormones and Environmental Signals

There is considerable evidence that abscisic acid (ABA) is an important positive regulator of both the induction of dormancy and the maintenance of the dormant state in imbibed seeds following shedding:

- ABA-deficiency during seed development is associated with absence of primary dormancy in the mature seed, whereas over-expression of ABA biosynthesis genes can increase seed ABA content and enhance seed dormancy or delay germination.
- ABA produced by the seed itself during seed development can impose a lasting dormancy, whereas, maternal ABA or ABA application during seed development both fail to induce a lasting seed dormancy, but has other functions.
- In *Arabidopsis thaliana,* the members of the AtNCED gene family encode enzymes catalyzing the key regulatory step in ABA biosynthesis. ABA synthesis in the embryo and the endosperm both seem to contribute to the induction of seed dormancy.
- High ABA contents are present in the imbibed seeds of the strongly dormant *A. thaliana* ecotype Cape Verde Island (Cvi) and decrease as dormancy is lost. A recent transcriptome analysis with this ecotype strongly supports the view that increased ABA biosynthesis is associated with the dormant state.
- *De novo* ABA biosynthesis during imbibition of dormant, but not non-dormant, seeds has been demonstrated in the *A. thaliana* ecotype Cvi, as well as in other species including *Nicotiana plumbaginifolia, Helianthus annuus,* and *Hordeum vulgare*. This de novo ABA biosynthesis has been interpreted as a mechanism for dormancy maintenance.
- Work with the strongly dormant *A. thaliana* ecotype Cvi shows that dormancy may depend on an intrinsic balance of gibberellin (GA)- and ABA-biosynthesis and catabolism, which will determine the dominance of either of the hormones. While PD release of Cvi seeds occurs effectively by after-ripening, stratification or inhibition of ABA biosynthesis, the addition of GA appears less effective. GA treatment of dormant Cvi seeds caused a transient increase in ABA levels, suggesting that in dormant seeds a feedback mechanism exists that

maintains a high ABA/GA ratio. Thus, the net result of the dormant state is characterized by increased ABA biosynthesis and GA degradation. It appears to be the ABA/GA ratio, and not the absolute hormone contents, that controls germination.

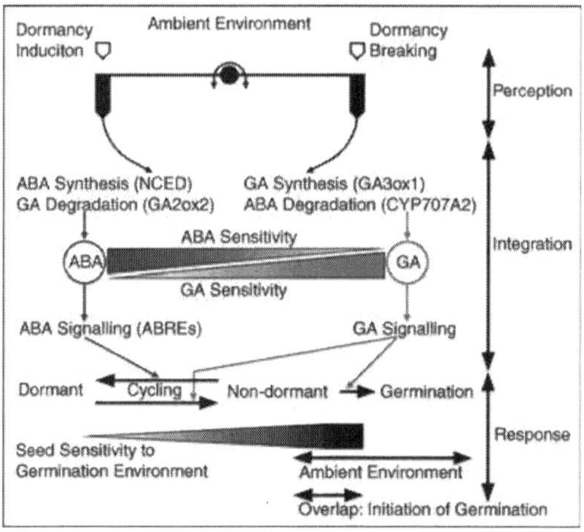

Model for the Regulation of Dormancy and Germination by ABA and GA in Response to the Environment

According to this model ambient environmental factors (*e.g.* temperature) affect the ABA/GA balance and the sensitivity to these hormones. ABA synthesis and signalling (GA catabolism) dominates the dormant state, whereas, GA synthesis and signalling (ABA catabolism) dominates the transition to germination.

The complex interplay between hormone synthesis, degradation and sensitivities in response to ambient environmental conditions can result in dormancy cycling. Change in the depth of dormancy alters the requirements for germination (sensitivity to the germination environment); when these overlap with changing ambient conditions, germination will proceed to completion.

EVOLUTION OF SEED STRUCTURE AND SEED DORMANCY

A general scheme of the evolutionary connection between seed structure and dormancy class.

MD and MPD are present in B-type seeds, but can also be typical of some specialized species. MD is thought to be the ancestral dormancy type among seed plants and is the most primitive dormancy class. The dispersal of seeds with an underdeveloped embryo that need time to grow might have evolved as an ancient strategy to disperse germination over time. MD and MPD are not only typical for primitive angiosperms, but also for primitive gymnosperms like the Zamiaceae, Ginkgoaceae, Podocarpaceae and Taxaceae. Evolution of larger embryo size in MD seeds resulted in non-dormant (ND) seeds. Concurrently, the gain of physiological dormancy mechanism(s) led from seeds with MD to seeds with MPD, which upon gain in embryo size led to PD seeds. PD is the most phylogenetically widespread dormancy class. This class of dormancy is distributed over the entire phylogenetic tree from gymnosperms and the basal angiosperms to the higher core eudicot Rosid clade. As ND is also distributed over the entire phylogenetic tree it has been proposed that gain and loss of PD occurred at several times and at several levels of seed evolution. The most phylogenetically restricted and derived dormancy classes are PY and PY+PD.

The occurrence of an impermeable seed or fruit coat combined with a ND embryo (PY) or a PD embryo (PY+PD) is probably the adaptation to specialised life strategies or habitats. PY and PY+PD are the only dormancy classes not found in gymnosperms, and PY is only evident in one Monocot family (Cannaceae). A conceptual model for the evolution of PY and PY+PD within the Ancardiaceae has been proposed. Within this family the genus *Rhus* includes species with PY+PD- dormancy (*e.g. Rhus aromatica*) and species with only PY-dormancy (*e.g. Rhus glabra*).

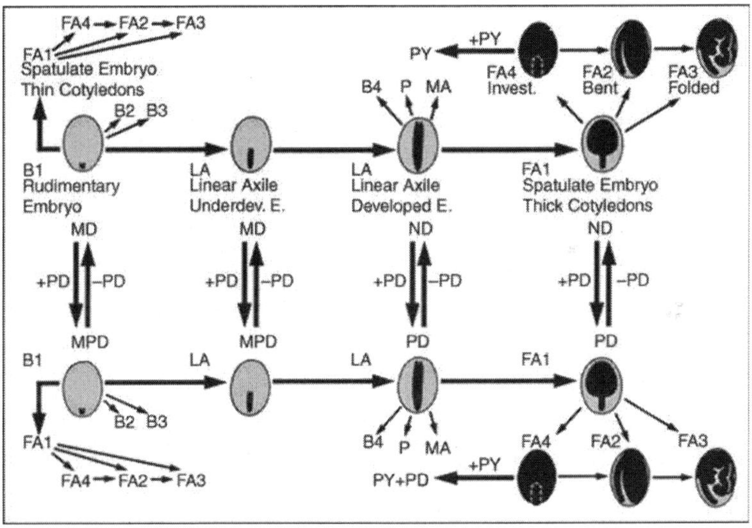

Anatomical evidence for PY of extinct *Rhus* spp. suggests that PY had evolved in the Ancardiaceae by 43 Ma in the Middle Eocene. The model begins in pre-Eocene times with an ancestral species that has large, non-dormant (ND), recalcitrant seeds. By the middle Eocene, a derived species with relatively small, orthodox seeds and a water-impermeable covering layer had evolved, *i.e.* a PY-dormant seed evolved. By the Oligocene, PD had been added to the seed dormancy mechanism, *i.e.* a PY-dormant seed evolved.

Climatic drying (Eocene), followed by climatic cooling (Eocene-Oligocene transition), are proposed as primary selective agents in the development of PY. An evolutionary connection between PY and recalcitrance is suggested by the relatively high concentration of these two states in the Rosid clade.

Phylogenetic data and fossil evidence seem to support the PY → PY+PD evolutionary sequence in Ancardiaceae, which also may have occured in Fabaceae.

Summary of General Trends in Angiosperm Seed Evolution with Respect to Seed Type and Dormancy Class

Seeds with small embryo and abundant endosperm (B1 seed type) are primitive and evolution shows a general trend (indicated by arrows) towards mature seeds with little or no endosperm in which the embryo occupies most of the seed (through LA to FA seed types). Gain or loss of PD to each of these seed types (B1, LA and FA1) alters the dormancy class as indicated.

12

Gymnosperms and Seed Dispersal

Saguaro seeds reach the ground in many ways. Some fall to the soil surface while they are still in the flesh of ripe, open fruits and in dried receptacles. Relatively heavy seedfalls occur during summer rainstorms. Rains soften and dislodge seeds from ripe fruits and dried receptacles remaining on the plant as well as from those that become lodged in the branches of associated trees and shrubs, and those already on the ground.

Summer rains aid importantly in transporting saguaro seeds to protected downslope sites affording both concealment from predators and favourable conditions for germination. After rains, seeds are found in the first few millimeters of the soil, in sand and gravel, under rock and in litter (usually mud-spattered), and nearly all are invisible to the naked eye. Natural germination occurs in such protected physical situations. Animals play a primary role in dispersal and the transport of seeds to sites suitable for germination and seedling establishment.

Most fruits drop to the ground, resulting in a high concentration of seeds at the base of the parent plant and providing an attractive supply of food for mammals, birds, and insects. The congregation of herbivorous predators1 not only results in low probabilities for seed survival but also severely limits seed germination and seedling establishment. Moreover, intensive digging, especially by rodents, effectively reduces protective plant cover at the base of the parent saguaro and destroys young saguaros or exposes them to other hazards. Generally, the probability of seed survival, germination, and seedling survival in creases with distance from adult plants. 1For lack of more appropriate terms we use the words "predator" and "predation" as broadly defined to include consumption and destruction of seeds and/or plants by animals.

Relatively low density populations of harvester ants (Pogonomyrmex sp.) and Harris' antelope ground squirrels (Citellus harrisi harrisi) are present, and round-tailed ground squirrels (Spermophilus tereticaudus) do not occur at this location. In the absence of heavy predation, a relatively large proportion of the annual seed cropremains on the ground until the occurrence of germinating rains.

Fig. Typical Absence of Fruit at the Base of Mature Saguaros in Flat Habitats at the East Monument. Detached Seed Masses are Absent and not Seeds are Present in 45 Receptacles Remaining on the Ground Beneath this Vigourous 5-branched, 11-m (36.1-ft) Tall Plant. One Ripe Fruit Still Remains on the Plant.

Relatively dense populations of harvester ants (Pogonomyrmex sp.) and round-tailed ground squirrels (Spermophilus tereticaudus) consume most of the seeds within a few hours after the fruits drop to the ground. In this habitat, the area surrounding the base of an adult saguaro provides little or no protection—either for seeds or seedlings—from the intensive foraging of mammals, birds, and insects. Photographed 19 July 1971.

In a matter of a few days, animal activity beneath the parent saguaro changes the situation from one resembling a fruit basket to the seed shadow associated with some tropical plant species as described by Janzen and Krebs. This explains why young saguaros are seldom found beneath the seed-producing parent plant.

Those few seedlings that begin their establishment there do not survive the continuing heavy animal traffic at such stations. In addition to providing an important mechanism for seed dispersal, birds frequently deposit seeds in protected locations provided by trees and shrubs away from areas of intensive animal foraging at the base of fruiting plants. Such protected locations not only increase the probability of seed survival, they generally offer more suitable conditions for seed germination and seedling establishment.

SEED CONSUMPTION AND LOSS

A large part of the annual saguaro seed crop is consumed by mammals, birds, and insects. The percentage of the annual seed crop so consumed varies greatly from one community to another depending upon the relative abundance of particular species of consumers.

Field observations in Saguaro National Monument indicate that the principal seed consumers are white-winged doves (Zenaida asiatica) and western mourning doves (Zenaida macroura) which feed on freshly opened fruits still attached to the plant, harvester ants (Pogonomyrmex sp.), and round-tailed and Harris' antelope ground squirrels (Spermophilus tereticaudus, Citellus harrisi) which feed intensively upon fruits that drop to the ground.

Fig. Galvanised Hardware Cloth Exclosure was Used in Experiments to Determine the Effect of Vertebrate Animals on the Survival of Saguaro Seeds and Young Plants at Saguaro National Monument. Net Germination in Plots so Protected was Approximately 1.3 times that in Adjacent Unprotected Plots. The Difference is Attributed Primarily to the Effect of Seed Removal by Birds and Rodents.

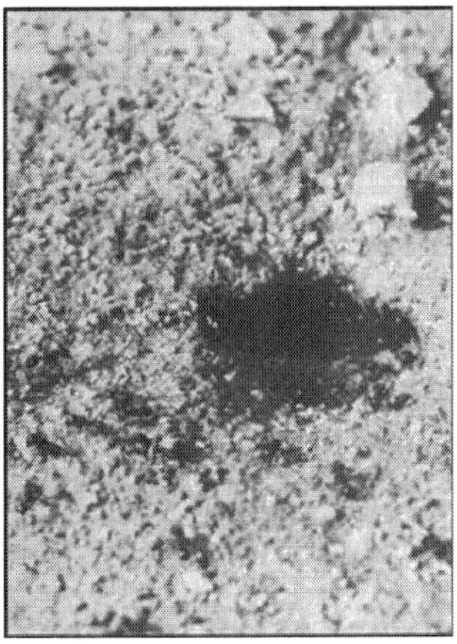

Fig. Harvester Ants (Pogonomyrmex Barbatus) Efficiently Remove Large Numbers of Saguaro Seeds that Fall to the Ground. In Localities such as the Cactus Forest of Saguaro National Monument (East) where they are Abundant, Harvester Ants play a Primary Role in Limiting the Number of Viable Seeds that Survive on the Ground Until the Occurrence of Summer Conditions Suitable for Germination.

Ordered by the estimated relative quantity of seeds consumed, other bird-consumers are Gambel's quail (Lophortyx gambeli), curve-billed thrasher, (Toxostoma curvirostre), cactus wren(Campylorhynchus brunneicapillus), gila woodpecker (Centurus uropygialis), gilded flicker(Colaptes chrysoides), and brown towhee (Pipilo fuscus).

Secondary mammal-consumers include the kangaroo rat, pocket mouse, white-throated woodrat (Neotoma albigula), cactus mouse(Peromyscus eremicus), long-nosed bat, coyote (Canis latrans), and collared peccary.

The proportion of the annual seed crop harvested by birds and mammals does not provide a direct measure of potential seed survival and natural germination. Rather, the impact of such consumption must be evaluated separately for each animal species based upon the diversity of its feeding habits and the efficiency of its digestive processes.

Seeds consumed by obligatory seed-eaters such as white-winged and mourning doves, kangaroo rats, pocket mice, and cactus mice are destroyed during digestion. However, a portion of saguaro seeds consumed by other species (including round-tailed ground squirrels, Harris' antelope ground squirrels, woodrats, coyotes, and omnivorous birds), which benefit primarily from the succulent tissues of the matrix, pass undamaged through their digestive tracts.

Coyote scats collected in July consisted of approximately 95 per cent undigested saguaro seeds. Viability is essentially unaffected by passage through the coyote digestive tract. Eddy reported collared peccary scats containing 85 per cent undigested saguaro seeds.

These mammal species together with the curve-billed thrasher, cactus wren, gila woodpecker, and gilded flicker are potentially important disseminators of saguaro seeds. Generally, the number and proportion of whole viable seeds present in mammal droppings increase with the size of the species and/or the diversity of its feeding habits. As with mammals, birds that are primarily seed-consumers are relatively low dispersers (viable seeds are not found in feces in large numbers).

In some localities, insects play a primary role in limiting the number of viable seeds that remain on the ground until conditions suitable for germination occur. Harvester ants efficiently remove large numbers of seeds. Other observed seed gathering taxa includeLasius sp. and Leptothorax sp.

The results of experiments utilising exclosures to measure the effect of predation on germination success are given in Table below. Low germination success (0-3.2 per cent) in plots at three sites was associated with and is attributed to seed removal by harvester ants.

Conversely, harvester ant activity was not observed in plots where comparatively high numbers of germinations were recorded. Low values for two of the east monument north-facing slope plots are attributed to the localised occurrence of relatively unfavourable soil characteristics associated with deep gravel.

The Effect of Predation on Germination Success in Protected and Exposed 0.25 m2(2.7 ft2) Plots. Plots were Seeded (1000 Seeds per Plot) on 26 June 1968 and Covered with 16 (0.16 cm; 0.06 inch) Mesh Window-screen (3 Replications), 0.5-inch (1.3 cm) Mesh Hardware-clsoth Exclosures, or Exposed (open) Until the First Rain (6 July) when all Plots were Covered by Window-screen Exclosures.

Plus Signs Indicate Observation of Harvester Ants(Pogonomyrmex) Activity. Site Locations are: SE = east Monument, SW = West Monument, F = Flats, RN = Rocky North-facing Slope, and RS = Rocky South-facing Slope

Table. The Effect of Predation on Germination Success in Protected and Exposed 0.25 m2(2.7 ft2) Plots.

Study site	Ants obs.	No. of seeds germinated (7/31-8/3)				Open plots		
		Exclosure plots					% Germ.	
		Screen A	Screen B	Screen C	Mesh		Excl.	Open
SERS	+	101	266	0	1	28	9.4	2.8
SWRS	+	0	0	0	66	112	11.6	11.2
SERN		30	239	31	35	13	8.4	1.3
SWRN		244	480	335	354	374	35.3	37.4
SEF		166	284	114	117	42	17.0	4.2
SWF	+	28	32	17	7	16	2.1	1.6
Total			2003		980	585		
Mean			111.3		163.3	97.5	12.4	9.8

The difference in germination percentages in open and in exclosure plots is a measure of the effect of seed removal by vertebrate consumers. Net germination within exclosure plots was 1.27 times that in open plots. The difference is attributed primarily to the elimination of seed predation by birds and rodents.

No harvester ant activity was observed on north-facing slopes. In other areas where harvester ants were present (south-facing slopes and flat habitats), their impact was highly localised. In some of these plots all visible seeds were removed within a few days after seeding, while seeds in plots a few meters distant were essentially undisturbed. It appears that high densities of harvester ant colonies are commonly associated with stands of reproductive (adult) saguaros.

Germination success in exclosures did not provide an effective means of comparing the separate effects of seed removal by birds, mammals, and insects. Rather, the results suggest that in competition with harvester ants, the impact of larger, less efficient seed-gatherers on seed survival is relatively unimportant; the principal effect is exerted by the most efficient forager.

Thus the high attrition of seeds between seedfall and germination is due to seed removal by animals. In the Tucson area, approximately 4 in 1000 saguaro seeds that reach a site where natural germination can occur survive to become seedlings.

During the 1- to 5-week pre-germination period, native animal predators (birds, mammals, and insects) are the principal agents of saguaro seed attrition.

Table. Summary of Natural Germination in Experimental Plots from Broadcast Saguaro Seeds, 1965-68.

Year	Sites (no.)	Plots (no.)	Seeds (1000)	Germinations	
				No.	%
1965	4	28	28	96	0.34
1966	4	13	36	87	0.24
1967	9	19	95	461	0.48
Subtotal	—	60	159	644	0.40
1968	8	40	40	7595	18.99

Experimental sites, all in representative saguaro habitats, were located at the east monument in 1965 and 1966, east and west monument in 1967 and, in addition, included one site on the south slopes of the Catalina Mountains in 1968.

The number of germinations shown are the total number of seedlings that sprouted between 1 July and 10 August in experimentally seeded study plots. Results for 1965-67 are from 60 unprotected plots exposed to all natural environmental factors. The mean annual germination rate in these plots was 0.353 per cent.

Greater germination success in 1968 reflects partial protection by exclosures used to limit animal predation activity. Generally, seeds remaining on the ground at summer's end are destroyed by other agents prior to the ripening of the next crop. Seemingly whole seeds from the previous year's crop collected in June from the soil surface proved to be nonviable. Seed coats showed signs of entry by small insects, or the embryos were withered and brown.

Seeds which remain over winter on the ground make no important contribution to the following summer's germinable seed supply. At Saguaro National Monument, the consumption of seeds by birds, mammals, and insects is higher at the east monument (Rincon Mountains) than at the west monument (Tucson Mountains).

An approximate measure of the relative seed attrition at these two locations is offered by July observations on seeds remaining on the ground beneath vigourous adult saguaros in equivalent stages of fruit maturation. In flat habitat at the west monument, 179.5 g (6.3 oz) of seeds (approximately 131,000 seeds) were recovered from fruits collected beneath a 9-m (29.5 ft) plant with three productive arms.

In similar habitat at the east monument, no seeds were present in 45 receptacles surrounding an 11-m (36.1 ft) plant with five productive arms. The difference in yearly seed loss between the east and west monument areas is attributed primarily to consumption by animals, the round-tailed ground squirrels—abundant at the east site and absent at the west site—and harvester ants present in much greater abundance at the east site.

SEED DISPERSAL IN GYMNOSPERMS

In the typical gymnosperm cone, ripening and drying of cone and seed causes the cone scales to open and release the seeds. Dispersal is by wind, assisted by the presence of seed wings in some genera *e.g. Pinus.* In some species of pine, the "closed-cone pines" *e.g. P. radiata,* there is usually an interval of months or years between ripening of cone and seed and the opening of the cone to release the seeds. In a few cases, such as the interior provenances of *Pinus contorta,* the cones open only when subjected to the heat of occasional fierce forest fires. On the other hand the cones of *Abies* and *Araucaria* disintegrate readily on the tree within a few weeks of ripening. Seed dispersal by animals is less common, but the "berries" of *Juniperus* and the fleshy fruits of *Podocarpus* are examples. In addition seeds of temperate conifers are collected and stored by rodents and some may germinate before being eaten.

PROCESS

At one extreme, certain species of mangrove are viviparous, the seeds germinating before they separate from the parent. At the other extreme, seed of some species may remain dormant but alive for many years, capable of germinating if an event occurs to break the dormancy state. The subject of dormancy is discussed later in this chapter. Between the viviparous and the deeply dormant seed types occur many types of seed which are capable of germination soon after seed shed provided that environmental conditions are suitable.

Just as fertilization initiates the transformation of the ovule into the ripe seed, so does germination transform the embryo within the seed into the independent seedling. For the purpose of laboratory testing, germination is defined as the emergence and development from the seed embryo of those essential structures which are indicative of the seed's capacity to produce a normal plant under favourable conditions.

At maturity and seed shed many seeds have lost the greater part of the moisture which they contained in earlier stages. For example the embryo and female gametophyte of *Pinus lambertiana* contain as much as 50 per cent moisture content (fresh weight basis) shortly after fertilization, but by the time of natural seed dispersal the moisture content of the embryo is reduced to 23 per cent and of the female gametophyte to 38 per cent. Reduced metabolic activity is associated with the drying of the seed, so that the embryo is in a temporarily resting or quiescent state, which in non-dormant seeds can be easily reactivated by suitable conditions.

These conditions are:

- Adequate moisture
- Favourable temperatures
- Adequate gas exchange and, for some species,
- Light.

There is considerable variation between species in the optimum levels of the different factors and there is frequently an interaction between them.

Germination consists of three overlapping processes:

1. Absorption of water mainly by imbibition, causing a swelling of the seed and eventual splitting of the seedcoat,
2. Enzymatic activity and increased respiration and assimilation rates which signal the use of stored food and translocation to growing regions,
3. Cell enlargement and divisions resulting in emergence of radicle and plumule.

In most seeds the radicle of the embryo is close to the micropyle, where absorption of water is easier and quicker than through the seedcoat. As the radicle swells, it exerts pressure on the seedcoat which commonly splits first at this point to free the radicle. This gives rise to the primary root which grows down into the soil and soon produces lateral roots. Subsequent stages depend on whether the species exhibits epigeal germination *e.g. Pinus*—the hypocotyl elongates and the cotyledons are lifted above ground - or hypogeal germination *e.g. Quercus* —the hypocotyl is undeveloped and the cotyledons remain on or in the ground. In hypogeal germination the cotyledons can have only a food storage function, or a haustorial function (in species in which food is stored in the endosperm *e.g.* palms, *Scorodocarpus),* while in epigeal germination they may also perform a valuable photosynthetic function during early growth of the seedling.

In epigeal germination anchoring of the young plant by the radicle is followed by rapid elongation of the hypocotyl which arches upwards above the soil surface and then straightens; simultaneously the cotyledons and plumule are exposed, to which the seedcoat may or may not still be attached. The plumule then develops into the primary shoot and photosynthetic leaves. In the subtype "durian germination" the hypocotyl elongates but the cotyledons are shed while still enclosed within the seedcoat. In hypogeal germination, the cotyledons remain *in situ* underground or on the ground while elongation takes place in the plumule.

In the subtype "semihypogeal germination" the cotyledons are exposed but remain on the ground. The two main types, epigeal and hypogeal and the two subtypes, durian and semihypogeal are the result of the four possible combinations of two independent variables: hypocotyl elongated or not and cotyledons exposed or not. All four combinations occur in the humid tropics.

Even in non-dormant seeds there is considerable variation, between species and individuals, in the speed of germination, from a few days to several weeks; much of this is due to different rates of imbibition in the first stage. Many tropical rain forest species, with high moisture content and permeable seedcoats at seed fall, must germinate within a few weeks. If they fail to find suitable conditions soon, they lose viability and die.

DORMANCY

The term "dormancy" refers to a condition in a viable seed which prevents it from germinating when supplied with the factors normally considered adequate for germination - suitable temperature, moisture and gaseous environment. A viable seed is defined as one which can germinate under favourable conditions, providing any dormancy that may be present is removed.

Dormancy in nature serves to protect the seed from conditions which are temporarily suitable for germination but which quickly revert to conditions too harsh for survival of the tender young seedling. Thus a seedcoat relatively impermeable to moisture prevents germination during isolated showers in the middle of a long dry season, while permitting it during a sustained rainy season. In the cool temperate zone the type of embryo dormancy which can be removed only through exposure to low temperatures facilitates subsequent germination in spring, while preventing it in autumn, when the resulting seedling would be unlikely to survive the winter.

The strength of dormancy has been observed to vary according to latitude and provenance, and from year to year even in seed from the same parent. There is also differential dormancy within the same species and seedlot, so that germination is staggered over a more or less extended period of time. In the Malaysian woody flora, about 50 per cent of species complete germination within six weeks or less, so that the differential is not very great, but in some species such as the hard-seeded leguminous *Parkia javanica* the germination period may extend from 1 week after sowing for the first seed to two years for the last.

Differential dormancy and staggered germination insures against the entire seed crop being destroyed by a single climatic catastrophe or pest. In nature a number of external factors may work, more or less slowly, to end seedcoat dormancy. These include alternate heating and cooling, alternate wetting and drying, fire and the activities of animals, soil organisms, fungi, termites and other insects. Dormancy due to embryo immaturity will come to an end if the embryo is given time and suitable conditions in which to mature after seed shed.

The exact mechanisms of physiological dormancy of the embryo, and of the processes which can terminate it, have been widely investigated but underlying causes are still little understood. There is good evidence that growth-promoting hormones, of which gibberellin is a well known example, and growth-inhibiting hormones interact in the maintenance or breaking of dormancy.

In temperate climates the balance between inhibitors and growth-promoters is altered by a combination of low temperature and high moisture maintained over a period of time which varies from species to species. This combination is provided naturally during winter, the season least suitable for growth. It can trigger biochemical changes in the embryo which lead to the breaking of dormancy, the initiation of embryo metabolism and growth, and the subsequent germination of the seedling.

Research on the physiology of tropical tree species has, unfortunately, been only a fraction of that done on temperate species. There is no reason to suppose that the combination of low temperature and high moisture—the technique of "stratification" when applied artificially—would have any effect on tropical seeds possessing embryo dormancy (if they exist). In the dry tropics, on the contrary, a combination of conditions typical of the season least favourable for growth - high temperature and low moisture—would seem more likely to trigger the breaking of embryo dormancy and lead to germination in the following rainy season.

In fact, seedcoat dormancy by itself appears to provide adequate protection for species of the dry tropics. From the forester's point of view dormancy has some disadvantages. Delayed and irregular germination in the nursery is a serious constraint on efficient nursery management. Much research has therefore gone into devising effective artificial treatments to remove dormancy, in order to ensure that the seeds germinate quickly and evenly in the nursery beds.

On the other hand, dormancy confers certain advantages. Not only does it improve the chances of survival in nature, as mentioned previously, but it preserves the seed against temporarily unsuitable conditions such as may occur during the period between seed collection and storage. High quality orthodox but non-dormant seeds, dried to the appropriate moisture content and stored at the correct temperature, should have as long a life in storage as dormant seeds, but dormancy provides an insurance against the loss of viability during transport and processing which can easily occur in non-dormant seeds in less than ideal conditions.

HAZARDS OF SEED PRODUCTION

Both the quantity and the quality of seed crops may be greatly affected by external factors. Climatic factors can affect the abundance of flowering and thus indirectly of seed production. There is some evidence that above-average temperatures and a modest degree of moisture stress in spring and early summer can induce abundant formation of flower buds in temperate regions. In Nigeria good seed years occur in *Triplochiton scleroxylon* following a particularly dry (30 per cent or less of average rainfall) August, the month when there is a diminution of rainfall between the heavy early and heavy late rains.

More extreme cases of unseasonal climate usually reduce the crop of flowers or fruits. Late spring frosts in temperate regions kill flowers or young fruits and abnormally high temperatures or drought may have a similar effect. Even if death and premature shed of whole fruits does not occur, a proportion of the seeds may abort later. Mechanical destruction of flowers or fruits may be caused by exceptionally high winds or hailstorms. Continuous rain at the time of pollen dispersal has a particularly adverse effect on the amount of seed set, whether pollination is by wind or insects. *Tectona* flowers during the rainy season and this may account for the low average rate of fertilization of 1–3 per cent reported from Thailand over the period 1967–72. Rain discourages flight of the pollinating insects as well as washing off pollen grains from the stigma before they germinate. Continuous wet weather during the pollen dispersal season is considered the main factor responsible for the characteristically poor seed crops of *Pinus merkusii* in Indonesia and Malaysia. Tamari reported that over 90 per cent of Dipterocarp flowers in Malaysia failed to develop into fruits.

Birds, mammals, insects, fungi and bacteria all do damage in flowering as well as fruiting stages. Insects are probably responsible for the most serious losses in the greatest number of species. For example, the weevil *Apion ghanaense* destroys a large part of the flowers and seeds of *Triplochiton* each year. The larvae of *Pagyda salvaris* may destroy as much as 90 per cent of the flower buds of *Tectona* in some years.

Two species of the Bruchid genus *Amblycerus* can destroy many seeds of *Cordia alliodora*, but damage can be reduced by collecting the seeds three weeks before natural seedfall. The weevil Nanophyes sp. may attack up to 60 per cent of the seeds of *Terminalia ivorensis.*

Cone worms of the genus Dioryctria have been known to damage 60 per cent of maturing cones and seeds of Pinus elliottii and P. palustris in the southern USA and the same genus can also cause severe damage on *Pinus merkusii* seeds in the Philippines. Larvae of *Agathiphaga,* a genus of moths, may destroy over 50 per cent of the seeds in cones of several species of *Agathis* in Queensland and the western Pacific islands.

Seeds of many dry area species of *Acacia* and *Prosopis* suffer serious damage from Bruchid larvae. Birds and mammals, especially squirrels, may consume considerable numbers of seeds in some years, although they also perform a useful service in seed dispersal. Losses from pests and diseases do not normally have a serious effect in years of abundant seed production, but in years when flowering is poor for climatic reasons they can convert a light crop into complete failure.

BEGINNING OF THE GERMINATION

Germination takes place only in the first week after the beginning of the summer rains, and the seedlings are hidden in the litter under bushes or among the small stones which cover the ground.

The saguaro is a subtropical species and its seeds germinate only in the summertime. Germination occurs principally in July and August during the southwestern summer monsoon. Thus, successful germination and early establishment depend on a short period of coincident high levels of moisture and warmth during but a few weeks of the desert year. We are not surprised, therefore, to find that this highly successful desert species has acquired adaptive strategies that fit a set of germination requirements to an environment characterised by a persistently high evaporation which is only partially offset by scanty and often uncertain rainfall.

Seedfall is completed just prior to the normal arrival of summer rains. In exceptional years, the two events—seedfall and rainfall—actually overlap. The first summer rains contribute importantly to seed survival by dispersing seeds and by washing them into locations that offer concealment from predators. Little or no germination, however, results from these late June and early July rains. Rigid light and moisture requirements interact to inhibit germination and prevent losses that would result from germination at that time, for sporadic early monsoon rains seldom provide a sufficiently reliable supply of moisture to promote germination and support the initial establishment of seedlings. The increasing frequency of rains during the second half of July offers progressively more favourable conditions for seed germination.

While it is commonly said that "seedlings germinate," plant seedlings "sprout" and proceed to establish or die; only seeds can germinate. The highly variable spatial and temporal distribution of summer rainfall importantly affects the number of saguaro seed germinations occurring at a given location. Further, the number of germinations varies in accord with the characteristically large year-to-year variability of monsoon precipitation patterns.

Our experimental and field observations, however, indicate that the most favourable conditions for saguaro seed germination do not necessarily occur during years with the highest total summer precipitation. Rather, natural germination is associated with temporal clustering of summer rainstorms that provide continuously high moisture levels at and near the soil surface during the 2- to 3-day germination period.

Most germination takes place during the period from mid-July through the first week of August. Then, with continuing availability of high surface moisture resulting from short-spaced rainstorms and supplemented by dewfall, germination proceeds to completion within a period of 2-3 days—initial establishment of the seedlings takes place at a time when temperatures and moisture are optimum for their growth.

The high temperature limitations on saguaro seed germination operate to restrict natural germination to shaded sites where periods of high moisture are sufficiently prolonged to permit germination. During wet periods such sites seldom reach temperatures high enough to inhibit germination. Inhibition of germination by low temperatures effectively restricts germination to the warm-wet periods of June, July, August, and September in the overall distribution of the saguaro.

Table. Per cent Soil Moisture in Shaded and Unshaded Saguaro-paloverde Habitat at Saguaro National Monument (east), July 1967

Date July	Soil moisture (%)			
	Shade		Open	
	0.5 inch (1.27 cm)	1 inch (2.54 cm)	0.5 inch (1.27 cm)	1 inch (2.54 cm)
18	15.23	15.34	6.01	8.12
19	10.06	9.63	3.13	6.69
20	3.40	8.87	2.36	5.45
21[a]	1.85	6.33	1.39	4.47
22	11.45	9.96	7.80	7.54
23	5.15	8.10	2.43	4.50

[a]Estimated by regression (log % on time).

Samples of the top 0.5 inch (1.27 cm) and top 1 inch (2.54 cm) of soil were collected at approximately 1300 hr beneath the crown of a mature foothill paloverde at 1 m (3.3 ft) from the trunk (shade), and from an unshaded (open) level site approximately S m (16.4 ft) outside the crown.

Seeds were broadcast beneath the crown of a foothill paloverde (Cercidium microphyllum) on 16 July immediately prior to receiving 0.20 inch (5 mm) precipitation the same day. No germination occurred during the period of high soil moisture following a heavy rain the next morning, followed by clear skies (17 July, 1.5 inches; 38 mm). Germination took place after two periods of light precipitation (0.04 inch; 1 mm, 0.21 inch; 5 mm) and broken overcast skies on 21 and 22 July. The first seedlings emerged on 23 July, 3 days after the start of these rains. Amount (bar height) and period (bar width) of precipitation is shown.

Saguaro seed germination is dependent upon the continued availability of moisture at the soil surface as determined by temporal distribution of rainfall, cloud cover, and other factors that aid in continuously maintaining high relative humidity and surface moisture for a period of 2-3 days during the summer rainfall season. Generally, natural germination is associated with the occurrence of two or more rains within a 2- to 3-day period. Genetically controlled variability in the time required for completion of the germination process further insures that the entire annual seed crop will not respond to conditions that are marginal for seedling establishment. Approximately 10 per cent of seeds tested under ambient summer light and temperatures required 4 days or more of continuous exposure to free water for completion of germination. Thus, even under conditions that permit extensive germination, a substantial portion of the germinable seeds remain viable and available for subsequent germination.

The light requirement may seem to present an apparent survival disadvantage because it seemingly dictates that germination must occur at or near the soil surface where the predation pressures are greatest and requisite moisture conditions are least apt to prevail. However, in view of the characteristic development pattern of the monsoon rains, an important adaptive strategy can be recognised. In the presence of insufficient moisture to complete germination and initial establishment of the seedling, the light requirement acts to stop the germination process without loss of seed viability. Buried seeds retain their viability, and will germinate quickly when exposed by subsequent rains, animal digging, other natural forces, or experimental exposure.

The light requirement insures germination sufficiently near the soil surface for the globular seedling to emerge and immediately receive adequate light for subsequent growth. Sensitivity to far-red light may also permit better utilisation of energy from nocturnal re-radiation thus hastening completion of the germination process. Field observations indicate that in nature germination of shallowly buried seeds accounts for a substantial portion of the annual saguaro seedling crop. This observation appears to conflict with knowledge that germinating seeds must receive exposure to light after they have imbibed water, that little or no germination (<1 per cent) will occur in total darkness. Extensive laboratory tests using a variety of environmental, mechanical, and chemical treatments have revealed no mechanism by which germination might occur naturally in the absence of light (Alcorn and Kurtz 1959; McDonough 1964). Based on successful use of acid treatments in promoting dark germination, McDonough (1964) suggested that passage through the digestive tracts of animals might promote germination in the absence of light. However, we have attained no success in the dark-germination of saguaro seeds recovered from the feces of native rodents or birds. In fact, the natural germination of buried saguaro seeds requires no complex mechanism for explanation. In view of the brief exposure and low intensities of light required for germination as found by all investigators, the natural translucence and gravelly surface of the soils of the saguaro habitat will admit sufficient light in most instances to permit germination of seeds buried to a depth of 5-10 mm (0.2-0.4 inches), the approximate maximum depth from which the seedling can emerge. A further mechanism for satisfying the light requirement exists in the soil churning and washing action occurring during typical summer rain storms. Such action permits burial of seeds after wetting and exposure to light.

Germination of saguaro seeds requires contact with free water and exposure to light (Alcorn and Kurtz 1959; Alcorn 1961a). Under ambient summer light and temperatures in a laboratory environment, initial germination of saguaro seeds maintained in continuous contact with free water occurs approximately 48 hr after initial wetting and 50 per cent germination is reached in approximately 72 hr. However, within the natural environment of the saguaro such prolonged periods (3 days) of continuously available water are a rare, almost non-existent occurrence during July and August when most natural germination takes place. Rather, natural germination commonly follows two or more shorter periods of water availability with interspersed drying conditions at the soil surface resulting from the occurrence of two or more distinct rainstorms within a 2- to 5-day period (Steenbergh and Lowe 1969). In a humid atmosphere, saguaro seeds hygroscopically imbibe and retain moisture required for germination. Such pre-exposure to high relative humidity effectively speeds the germination process and reduces the required period of seed contact with free water. First germination of seeds so pre-conditioned can take place after 24 hr of wet contact and 50 per cent germination can occur within 48 hr. In a near-saturated atmosphere, maximum hygroscopic imbibition (approximately 20 per cent of the air-dry seed weight) is reached in approximately 20 hr.

Table. Hygroscopic Imbibition by Saguaro Seeds in the Laboratory at High Relative Humidities (90-100 per cent).

Time (hr)	Weight (g)	Uptake (%)
I - Ambient Light		
0	2.00	0.0
1	2.10	5.0
2	2.16	8.0
4	2.23	11.5
6	2.26	13.0
17	2.35	17.5
19	2.36	18.0
21	2.32	16.0
23	2.30	15.0
25	2.29	14.5
II - Total Darkness		
0	2.00	0.0
12	2.36	17.8
15	2.37	18.5
21	2.41	20.5
23	2.41	20.5
25	2.40	20.0

Air-dry 2-g (0.07-oz) seed samples were placed in petri dishes suspended over saturated paper towels in closed chambers in full room shade (indirect light) and ambient July air temperatures (ca. 70-95°F; 21-35°C) Data are given as weight and as moisture uptake (per cent air-dry weight). Dashed line represents break in curve at maximum moisture uptake. Sample I (in clear plastic chamber) was subjected to ambient light; sample II (in a light-proof metal chamber) received light only at the times of weighing. In a near-saturated atmosphere without condensation, saguaro seeds absorb atmospheric moisture up to approximately 20 per cent of their air-dry weight. Maximum hygroscopic imbibition is reached in approximately 20 hr, *i.e.*, ca.

one day. Data in Table below. In an experiment to determine the effect of such treatment on the required period of contact with free water, treated seeds reached 50 per cent germination in 46.5 hr, 21.5 hr sooner than untreated (air-dry) seeds (68.0 hr). Hygroscopic imbibition can reduce by approximately one day the required pre-germination period of seed contact with free water. Number and per cent germination of humidified (pre-treated) and of air-dry (untreated) saguaro seeds in the laboratory under ambient summer light and temperatures (full shade, ca. 65-95°F; 18.3-35.0°C) Pretreated seeds were held 5 days in open petri dishes in a near saturated atmosphere immediately prior to testing. Hours to germination of treated (2—1.00 seed lots) and untreated (3—10.0 seed lots) seeds is measured from time of initial contact with free water (15 August, 0800 hr).

Hours Wet	Germination			
	Humidified (20.0 Seeds)		Air Dry (300 Seeds)	
	No.	%	No.	%
24	7	3.5		
26	16	8.0		
28	27	13.5		
30	39	19.5		
42	95	47.5	11	3.7
48	116	58.0	22	7.3
54	148	74.0	88	29.3
72	167	83.5	169	56.3
120	184	92.0	254	84.0

Hygroscopic imbibition can reduce by approximately one day the required period of seed contact with free water to germinate under natural conditions. Thus, in the desert environment where moisture is critically limiting, chances for germination are greatly increased by this adaptive mechanism that can reduce—by as much as a whole day—the required period of contact with a saturated soil surface. Germination 50 per cent points determined by regression analysis after probit transformation. Germination under natural conditions, therefore, need not depend upon the rare occurrence of a continuously saturated soil surface over a prolonged period of time; but rather it can be sustained by a period of high relative humidity followed by a relatively brief period of saturation. Thus, hygroscopic imbibition allows rapid germination with intermittent high moisture levels commonly associated with the characteristic summer storm patterns of the Sonoran Desert. Requisite environmental conditions for saguaro seed germination occur far beyond the bounds of the plant's natural distribution. Only at the western boundaries of its occurrence—in the lower Colorado River Valley—where summer rainfall is seldom adequate to promote germination, does the range of the species appear to be limited primarily by germination requirements. With that exception, therefore, the factors that limit its distribution and control populations along the margins of its range must act during the post-germination stage of the plant's development.

DISPERSAL RESULT

Early summer rains in late June and early July effect a major contribution to the dispersal of saguaro seeds by dislodging seed masses still remaining on the plant, and by washing individual seeds into protected locations that provide concealment from predators and are favourable sites for subsequent germination and establishment. The early rains reduce the attrition of seeds and ordinarily set the stage for germination during subsequent summer rains. Such seed dispersal to more favourable sites accomplished by gravity and rainfall run-off is mainly effective for downslope dispersal in sloping habitats. Animal consumers, principally birds, which pass undigested viable seeds through their digestive tracts, are the primary agents for seed dispersal upslope and in flat habitats. Obligatory seed-eaters that efficiently digest consumed seeds, however, make little or no contribution to seed dispersal. Attrition. In the complete life cycle of the saguaro, the greatest mortality occurs during the pre-germination (seed) stage, the 1- to 5-week period between seedfall and the occurrence of germinating summer rains. During that period the major portion of the annual seed crop is consumed by animals—birds, mammals, and insects. To some degree, the fruits of the saguaro are utilised either as food or a source of moisture by nearly every warm-blooded animal member of the community and by several species of insects. Initially heavy losses to feeding birds, especially doves, occur while fruits are still attached to the plant. The greatest losses, however, accrue after fruits drop to the ground where the diversity of consumer species ranges from ants to coyotes. The relative importance of consumer species varies according to their size and foraging efficiency, diversity of feeding habits, and relative abundance within the particular community.

Generally, obligatory seed-eaters—principally harvester ants, doves, and heteromyid rodents—exert the greatest impact as they are abundant and efficient consumers. However, in habitats where they are abundant (such as the Cactus Forest of the east monument), round-tailed ground squirrels consume the major portion of dropped fruits. The impact of specific saguaro fruit consumers varies greatly from one desert community to another. In flat habitats of the east monument, relatively dense populations of harvester ants and round-tailed ground squirrels quickly remove most of the seeds which reach the ground. However, in flat habitats of the west monument where neither species is abundant, a large proportion of the seed crop remains on the ground, undisturbed throughout the pre-germination period. In some locations, seeds remain over winter on the ground. However, as a result of destruction by insects, microorganisms, or climatic action, few if any survive or contribute significantly to the following year's germinable seed supply. Most of the fruits drop to the ground immediately beneath the parent plant. Intensive animal activity sponsored by this abundance of food not only results in heavy attrition of seeds from these sites but also severely limits the suitability of such sites for germination and seedling survival, especially in flat habitats. In view of abundant germination observed during favourable years, it is unlikely that the lack of young saguaros evident in some habitats is attributable to attrition of seeds.

Germination: The principal germination of saguaro seeds takes place in July and August from seeds of the current year's crop. Optimum conditions for natural germination coincide with the first full development of monsoon storms. Then, during the period from mid-July to mid-August, germination is normally associated with the occurrence of two or more rainstorms within a 2- to 5-day period. The availability of moisture is the critical determinant in the germination process; temperature and light requirements are readily satisfied within all natural habitats of the saguaro. In continuous contact with free water, germination takes place in 48-72 hr.

The germination process is facilitated by the capability of seeds to absorb moisture hygroscopically. We conclude that this is an important adaptive strategy. Pre-conditioning by hygroscopic imbibition can reduce the critical period of required contact with free water by as much as one day, importantly mitigating the need for a prolonged period of saturation at the soil surface. The saguaro is dependent in its germination requirements on a physically modified microenvironment produced by trees, shrubs, rocks, or other shade-producing objects. These shaded microhabitats provide moderated daytime temperatures that are within the upper range required for germination and prolong periods of high moisture availability at the soil surface. Ultimately, the number of adequately shaded sites suitable for saguaro seed germination is limited by the physical structure of the community.

Experimental evidence indicates that fewer than 1 in 200 seeds that reach a site where germination can occur survive to the seedling stage. We estimate that the net natural survival to the initial stage of seedling establishment is less than 1 per 1000 seeds produced. The summer climatic environment of the Tucson area is well within the range necessary for saguaro seed germination; this exceeds, in germination suitability, the drier, more westerly portions of the species' range. At Tucson, depending upon the overall intensity of monsoon development, the year-to-year suitability of conditions for natural germination ranges from poor to near optimum, but some natural germination occurs in all years. It seems likely that the western limits of saguaro distribution in California, Arizona, and northern Sonora are controlled by insufficient moisture for germination. Elsewhere to the northeast and south, it is clear that summer climatic environments would permit natural germination far beyond the present limits of saguaro distribution. Those limits, therefore, must be determined by factors that operate during the post germination stages of the plant's growth.

Index